T0296583

EXAMPLES IN HEAT AND HEAT ENGINES

EXAMPLES IN HEAT AND HEAT ENGINES

BY

T. PEEL, M.A.

FELLOW AND LECTURER OF MAGDALENE COLLEGE, CAMBRIDGE
HOPKINSON LECTURER IN APPLIED THERMODYNAMICS

CAMBRIDGE
AT THE UNIVERSITY PRESS
1935

CAMBRIDGE
UNIVERSITY PRESS

University Printing House, Cambridge CB2 8BS, United Kingdom

Cambridge University Press is part of the University of Cambridge.

It furthers the University's mission by disseminating knowledge in the pursuit of education, learning and research at the highest international levels of excellence.

www.cambridge.org
Information on this title: www.cambridge.org/9781316633298

First edition 1919
Reprinted 1922
Second edition 1935
First paperback edition 2016

A catalogue record for this publication is available from the British Library

ISBN 978-1-316-63329-8 Paperback

PREFACE TO FIRST EDITION

The questions in the following collection have with few exceptions been taken from papers set for the courses of lectures at the Cambridge Engineering Laboratory, and from the A and B papers of the Engineering Tripos. For the convenience of students they are arranged in short papers increasing in difficulty and in the variety of the subjects. Questions upon the dynamics of heat engines have not been included.

My thanks are due to Mr E. P. Moullin, of Downing College, for his kindness in checking a large number of the answers.

T. PEEL

CAMBRIDGE
October 1919

PREFACE TO SECOND EDITION

The book has been enlarged and revised and the questions, Papers I to XL, arranged under headings, indicating the various parts of the subject. In Papers XLI to L the questions are varied and of a general character. They do not go beyond the standard of the A papers of the Tripos.

I am very grateful to Mr Womersley for his kindness in checking the answers.

T. PEEL

CAMBRIDGE
October 1935

CONTENTS

PAPER I

1. 1000 gallons of water are being pumped per min. to a height of 80 ft. The pumps have an efficiency of 58 per cent. and are driven by a steam engine. Find the I.H.P. of the engine if its mechanical efficiency is 78 per cent. The kinetic energy of the water at delivery may be neglected.

If the engine requires 2 lb. of coal per I.H.P. hr., and the value of the coal is 7500 C.TH.U. per lb., what is the efficiency of the whole plant?

2. In the experiments of Osborne Reynolds to determine the mechanical equivalent of heat, a paddle was fixed to the shaft of an engine and rotated in water within a closed hollow vessel mounted freely on the shaft and prevented from turning round by weights attached to its side. At 300 R.P.M. of the engine 470 lb. of water were found to have a rise of 99·5° C. in 62 min. Determine the mechanical equivalent of heat if the load on the side of the vessel is found to be 140 lb. at a leverage of 4 ft.

3. In the manufacture of lead pipes the solid lead is squirted through an annular die by pressure on a plunger working in the cylinder in which the lead is contained. If the pressure necessary to force the lead through the die be 20,000 lb. per sq. in., shew that the lead just after leaving the die will be about 90° C. hotter than in the press.

Density of lead = 11·5. Specific heat of lead = 0·032.

4. An iron wire is suddenly loaded to a stress of 40,000 lb. per sq. in. and stretches under this load by $\frac{1}{20}$ of its length. Shew that its temperature rises by about 4° C. The weight of iron is 480 lb. per cu. ft.

5. An iron bullet is fired obliquely at a hard steel plate with a velocity of 1600 ft. per sec. The bullet is deformed by the impact but is not broken, and its velocity after striking is 800 ft. per sec. The plate is unaffected. If the temperature of the bullet

before striking is 20° C., shew that its temperature after striking is about 190° C. The average specific heat of iron is 0·12 for the range 20° to 100°, and 0·13 for the range 100° to 200°.

6. 18,000 gallons of water are being pumped per hour to a height of 70 ft. by a pump driven by an oil engine. The pump has an efficiency of 66 per cent. Determine the I.H.P. of the oil engine if its mechanical efficiency is 75 per cent. The kinetic energy of the water at delivery may be neglected.

If the oil is of specific gravity 0·8 and the engine requires 0·1 gallon of oil per I.H.P. hr., what is the efficiency of the whole plant if the calorific value of the oil is 10,000 TH.U. per lb.?

7. A gas engine driving a dynamo takes 520 cu. ft. of gas per hour when the dynamo is delivering 120 amperes at 110 volts. The calorific value of the gas is 320 TH.U. per cu. ft. Find the efficiency of the transformation of thermal into electrical energy.

If the output of the dynamo is 72 per cent. of the indicated work in the cylinder, find the rate of consumption of gas in cu. ft. per I.H.P. per hr.

8. To determine the mechanical equivalent of heat an electric motor was coupled to a Froude brake. With the motor giving a small output the brake load was found to be 11·25 lb. at 2000 R.P.M. and the flow of water 3·1 lb. per min., raised in temperature from 14·1° C. to 49·6° C.

The motor was then put on full load and the brake and the flow of water were adjusted until the speed and the inlet and outlet temperatures were the same as before. Under these conditions the brake load was 135 lb. and the flow 39·4 lb. per min.

The H.P. absorbed by the brake was $\frac{WN}{4500}$, where W was the brake load in lb. and N the speed in R.P.M.

Calculate the value of the mechanical equivalent.

What correction is made by using two sets of readings as above?

PAPER II

1. An air thermometer graduated at the pressure of 760 mm. of mercury is used to determine a temperature when the barometer stands at 30·95 in.: determine the true temperature when the reading of the thermometer is 51° C.

2. A balloon of 5000 cu. ft. capacity is to be so far filled with hydrogen at 30 in. of mercury and 15° C. that after ascending to a height where the pressure is 20 in. of mercury and the temperature 0° C., the silk envelope is then fully extended, no gas having been spilled. Calculate the mass of hydrogen required and its original volume.

3. A bicycle pump which has a stroke of 8 in. is used to force air into a tyre against a pressure of 50 lb. per sq. in. by gauge. What length of the stroke of the pump will be swept through before air begins to enter the tyre if the piston is pushed in (i) slowly, (ii) quickly? The barometric pressure is 15 lb. per sq. in.

4. A quantity of air at a temperature of 20° C. and pressure 30 lb. per sq. in. is found to be of volume 6·49 cu. ft. The volume is kept constant while heat energy equivalent to 10,000 ft.-lb. is given to the air: find its new pressure and temperature.

5. 10 cu. ft. of air of temperature 50° C. and pressure 100 lb. per sq. in. is allowed to expand to three times its volume: find the resulting pressure and temperature when the law of expansion is (i) $pV = \text{constant}$, (ii) $pV^{1\cdot2} = \text{constant}$, (iii) $pV^{0\cdot8} = \text{constant}$.

Determine the amounts of heat taken or rejected by the gas in the three cases.

6. One pound of a gas is expanding adiabatically and its temperature is observed to fall from 240° C. to 116° C. while the volume is doubled. The gas does 30,000 ft.-lb. of work in the process. Find the values of the specific heats k_p and k_v.

7. A quantity of air weighing $\frac{1}{4}$ lb. expands from a volume of 1 cu. ft. and pressure of 100 lb. per sq. in. to a volume of 10 cu. ft. at 20 lb. per sq. in., the pressure falling uniformly: find the heat taken in by the air.

8. Find the rate of loss of heat with rise of temperature when a perfect gas is being compressed according to the law $pV^n =$ const. $(n<\gamma)$.

For a Diesel engine, the charge may be taken essentially as air $(\gamma=1{\cdot}40)$. Shew that with $n=1{\cdot}35$, the working fluid on the compression stroke is losing heat at the rate of about 34 ft.-lb. per degree C. per lb.

PAPER III

1. A balloon of 5000 cu. ft. capacity is filled from gas cylinders at a pressure of 500 lb. per sq. in. The balloon is filled very slowly and the cylinders are reduced to 20 lb. per sq. in. in pressure before the balloon is full. Determine the volume of the cylinders and the heat taken from the atmosphere.

2. One pound of air is at 100 lb. per sq. in. and 100° C. Find its volume, its internal energy, and its entropy, reckoning the internal energy and entropy as zero in the standard state.

The air is allowed to expand to four times its initial volume (1) at constant pressure, (2) isothermally, (3) adiabatically. Find the final values, in each case, of the pressure, the temperature, the internal energy, and the entropy.

3. Find the rate of rejection of heat with rise of pressure $\left(\frac{dQ}{dp}\right)$ when a perfect gas is compressed according to the law $pV^n =$ const.

In a single stage air compressor the I.H.P. is 5 and the compression is according to the law $pV^{1{\cdot}25} =$ const. Shew that about 35·4 TH.U. per min. are rejected from the air during compression. Suction and delivery are at constant pressure and clearance is neglected.

4. One pound of air at a temperature of 100° C. and pressure 50 lb. per sq. in. is found to be of volume 4·97 cu. ft. The pressure is kept constant while heat energy equivalent to 5000 ft.-lb. is given to the air: find its new volume and temperature.

5. If air expands according to the law $pV^{1{\cdot}2} =$ const., shew that the heat supplied during the expansion is equal to the loss of internal energy, and that the rate of gain of entropy with

respect to temperature is inversely proportional to the absolute temperature.

6. Prove that when a gas expands according to the law $pV^n = \text{const.}$ the rate of heat reception with change of volume at any instant is given by

$$\frac{\gamma - n}{\gamma - 1}\, p,$$

where p is the absolute pressure.

A single stage air compressor is compressing air from normal atmospheric pressure to five times its initial density and delivering it at a pressure of 111 lb. per sq. in. absolute. The horse power absorbed in driving the compressor is 5 and its mechanical efficiency is 0·85. Shew that the heat lost through the cylinder walls of the compressor is about 29 C.TH.U. per min. γ for air may be taken as equal to 1·408.

7. On the expansion curve of an oil engine it was found that the index n had a value 1·18 between temperatures of 1190° C. and 1060° C. Shew that about 7 per cent. of the heat value of the fuel is accounted for by after-burning over this period, if the engine takes in 0·1 lb. of air per cycle and the oil supplies 40 TH.U. (Assume that the working fluid has the properties of air and that heat loss can be neglected.)

8. A movable piston in a closed rigid cylinder is held separating two quantities of a gas of equal volume V and of equal temperatures but of unequal pressures p_1 and p_2. Shew that if the piston is released the resulting pressure of the gases when in equilibrium is

$$\frac{1}{2}\left(p_1 + p_2 + \frac{(\gamma - 1)\, J . Q}{V}\right),$$

where Q is the number of heat units that have leaked into the gases.

PAPER IV

1. One cu. ft. of air initially at 15° C. and at a pressure of 100 lb. per sq. in. is heated at constant pressure to a temperature of 200° C. and then expands adiabatically to a temperature of

100° C. when it is cooled at constant volume to 15° C. and restored to its initial volume and temperature by an isothermal compression.

Estimate the heat taken in, the work done, and efficiency.

2. Air at a temperature of 100° C. expands at constant pressure to a temperature of 180° C. and then falls in pressure at constant volume until an adiabatic compression of the air to its original volume will restore it to its original pressure. Find the efficiency of this cycle.

3. Air at a temperature of 19° C. and at a pressure of 15 lb. per sq. in. abs., the volume of 1 lb. being 13 cu. ft., is compressed to a pressure of 1200 lb. per sq. in., the law of compression being $pV^{1 \cdot 1}=$const. It is then allowed to cool, at constant volume, to its original temperature, and afterwards does work in an air motor, the law of expansion being the same as the law of compression, and release taking place at a pressure of 25 lb. per sq. in. absolute. Draw the p-V diagram for these operations, and find the work done (1) on each pound of air in compression, (2) by each pound of air in expanding.

4. A cubic foot of air at atmospheric pressure of 15 lb. per sq. in. and temperature of 100° C. is compressed adiabatically in a cylinder until its pressure is 500 lb. per sq. in.: it is then heated at constant pressure until its temperature is 2300° C. and afterwards expanded adiabatically to its original volume and at this volume falls to its original pressure. Find the work done by the air in this cycle and determine the efficiency of the cycle.

5. A cylinder provided with a piston contains 1 lb. of dry air at 273° C. and a pressure of 367½ lb. The air expands adiabatically to 5 times its original volume, it is then compressed isothermally to its original volume, and the cycle is completed by supplying heat at constant volume until the initial conditions are obtained. Find (1) the work done in ft.-lb., (2) the heat taken in and rejected, (3) the thermal efficiency of the operation.

6. 1 cu. ft. of air at a temperature of 387° C. and pressure 100 lb. per sq. in. expands isothermally to a volume of 5 cu. ft., and at this volume is cooled, falling in pressure until an adiabatic compression of the air will restore it to its original pressure when

at its original volume 1 cu. ft. Determine the heat taken by the air, and the heat rejected in this cycle.

Draw the τ-ϕ diagram for the cycle and calculate the changes of entropy.

7. 3 cu. ft. of air at a temperature of 50° C. and pressure of 100 lb. per sq. in. abs. is heated at constant pressure until its volume is 12 cu. ft.: it is then cooled at constant volume until its temperature is again 50° C., and compressed isothermally to its original volume. Determine the change of entropy in each stage of the cycle; find also the work done and the efficiency of the cycle.

8. A Stirling engine with perfect regenerator works between temperatures of 16° C. and 305° C. The lowest pressure in the cycle is 15 lb. per sq. in., and the highest is 120 lb. per sq. in. Find (1) the thermal efficiency, (2) the ratio of expansion and compression.

Find also the thermal efficiency if the efficiency of the regenerator is 40 per cent.

The engine has a stroke of 18 in. and develops 12 H.P. when running at 50 R.P.M. What is its piston diameter?

PAPER V

1. 1 lb. of air is taken round (1) a Carnot cycle, (2) a Stirling cycle. In each case the highest pressure is 100 lb. per sq. in. and the lowest pressure is 10 lb. per sq. in. The highest temperature is 200° C. and the lowest temperature is 25° C. Find the p-V co-ordinates of the four corners of the diagram for each cycle and tabulate for each stage the values of Q, the heat supplied; W, the external work done; and E, the internal energy added.

2. A body at temperature 200° C. can give out 500 TH.U. in cooling to 20° C. If 20° C. is the lowest temperature available, shew that the greatest amount of work that can be done is 154,000 ft.-lb.

3. Find the maximum amount of work that can theoretically be obtained from 1 lb. of water at 100° C. if all the heat wasted may be given out at 15° C. Assume the specific heat of water to be constant and equal to unity.

4. How much work, in ft.-lb., can theoretically be done, in accordance with the theory of Carnot's engine, by a passage of heat from 1 lb. of water at 150° C. to 1 lb. of water at 50° C., which results in equalising their temperatures? What is the common temperature obtained?

5. The specific heat of a substance at an absolute temperature T may be taken to be given by $a+bT$, where a and b are constants. Shew that if the heat from unit weight of the substance be used as efficiently as possible in doing work, the quantity of heat wasted will be equal to

$$T_2 \left\{ b\,(T_1 - T_2) + a \log_e \frac{T_1}{T_2} \right\},$$

T_1 being the initial temperature of the substance and T_2 the lowest available temperature.

6. A heat engine draws its supply of heat from a source at temperature 260° C. but takes in only half its supply at this temperature, the other half being taken in at temperatures gradually falling from 260° to 150° C., the supply being uniformly distributed over this range of temperature. The heat rejected is all given out at a temperature of 40° C. What is the maximum efficiency of this engine and what fraction is it of the maximum efficiency for the given range of temperature?

7. A heat engine takes in 500 TH.U. of heat at a temperature of 150° C. and rejects its heat to a body whose temperature rises 1° C. for every 5 TH.U. of heat absorbed. If the temperature of the body is initially 15° C., shew that the maximum work which the engine can do is 115 TH.U. and that the final temperature of the body will then be 92° C.

8. An engine A working a reversed heat engine R is to be employed for the purpose of warming a building: the actual thermal efficiency of engine A is one-eighth. Engine R takes heat at a temperature of 0° C. and rejects it at 25° C. If working forwards, between the same limits of temperature, its relative efficiency may be expected to be six-tenths. It may also be assumed that 15 per cent. of the energy exerted by A is expended in overcoming the friction of the mechanisms of the two engines.

Suppose the heat available for warming the building to consist of all the heat rejected by R, and three-fourths of that rejected by A: determine the ratio of this quantity to the heat expended on engine A.

PAPER VI

1. If dQ represents a quantity of energy supplied thermally to unit mass of a perfect gas at temperature T, shew that

$$\int_1^2 \frac{dQ}{T} = K_p \log_e \frac{V_2}{V_1} + K_V \log_e \frac{p_2}{p_1},$$

and calculate the value of this when 1 lb. of air is compressed from a pressure of 16 lb. per sq. in. and temperature 16° C. to a pressure of 120 lb. per sq. in. and temperature 100° C.

2. A heat engine is working between 250° C. and 50° C. Of each 100 units of heat supplied to it 20 are taken in uniformly by the working substance as its temperature rises from 50° C. to 250° C., its specific heat being constant. Of the 80 units supplied at the high temperature 10 pass by direct conduction to the low temperature receiver. Find the maximum efficiency of the engine and compare it with that of a Carnot engine working between the same limits of temperature.

3. A building is to be heated by passing the air for ventilating it through a compressor which compresses it adiabatically then throttling it down to atmospheric pressure on leaving the compressor. The air enters the compressor at 0° C. and leaves it at 25° C. and the flow of air is 80 lb. per min. Calculate the power required, assuming unit mechanical efficiency.

If the power thus usefully employed cost 3*d.* per kilowatt hour, compare the cost of heating thus with that when burning coal in a stove the iron chimney of which passes through the air duct and imparts 20 per cent. of the heat of combustion to the ventilating air. Cost of coal 0·1*d.* per lb., and calorific value 8000.

4. A working substance performs a cycle in which it takes in energy thermally at a constant rate as its temperature rises from 160° C. to 200° C., its specific heat being constant and equal to 0·24. It rejects only at a temperature of 20° C. What is the maximum efficiency of the cycle?

The source of supply is maintained at a temperature of 210° C. and the heat rejected passes to a body at a temperature of 15° C. What increase of entropy takes place in the system as a whole, for each unit mass of the working substance which completes the cycle?

5. A heat engine receives 30 per cent. of its heat supply uniformly with the rise of temperature as the temperature of the working substance rises from T_2 to T_1, 60 per cent. at the upper temperature T_1 and 10 per cent. uniformly with the fall of temperature as the temperature falls from T_1 to T_2 again. All unused heat is rejected at T_2. Determine the efficiency of the engine if T_2 is one half T_1.

6. In a Carnot cycle the temperature of the source of heat is 200° C. and of the receiver of heat 15° C. Suppose that, in the portions of the cycle in which the working substance gains heat from the source and loses it to the receiver, plates of bad-conducting material are interposed between the cylinder and the source and receiver respectively, so as to introduce a temperature difference of 30° in the thickness of each plate. Write down in tabular form the changes of entropy in the source, receiver, working substance and whole system, in each step of the cycle of operations, if unit heat is taken from the source.

7. A cooling chamber is maintained at a temperature of 0° C. by an air engine working on a reversed Carnot cycle, the higher temperature being that of the atmosphere. Water passes through a coil of pipe, placed in the chamber, entering at atmospheric temperature and leaving at 5° C., and 1000 gallons of water are cooled per hour. Find the horse power required to drive the engine, neglecting mechanical losses and heat leakage (1) when the atmospheric temperature is 16° C., (2) when it is 26° C.

Find also the mass of working air in each case, if the ratio of isothermal expansion is 2 to 1, and the engine is running at the rate of 100 cycles per min.

8. *A* and *B* are two air reservoirs; *A* containing air at high pressure and temperature, and *B* containing air at comparatively low pressure and temperature. *A* supplies air to an engine *C* working adiabatically and exhausting without drop of pressure into *B*; and air is taken at the same rate from *B* and compressed

adiabatically and forced back into A by an air compressor which is driven partly by C and partly by external power. A has cooling surfaces and B heating surfaces, each uniformly supplied with water, one supply thus being warmed and the other cooled. The apparatus is supposed to be working steadily, A and B being large enough for this to be a possible assumption. It is found that every minute 10 lb. of water is cooled from 10° to 5° C., and 3 lb. of water is warmed from 10° to 45° C. Find (neglecting frictional losses) the external horse power employed, and the ratio of expansion in C.

PAPER VII

1. In a condenser trial the exhaust steam has a temperature of 55° C. at entry, and 15 lb. of circulating water are used for each pound of steam condensed. The circulating water has its temperature raised from 13° C. to 33° C., and the temperature of the air-pump discharge is 35° C. Determine the dryness of the exhaust steam supplied to the condenser.

2. Steam expands according to the law $pV = \text{const.}$, from a pressure of 160 lb. per sq. in. to a pressure of 2 lb. per sq. in. and it is just dry at the end of the expansion. Calculate the dryness fraction at the point of cut off and the heat received or rejected per lb. during expansion.

3. Draw up a table of the steam volume, temperature, water entropy and steam entropy of 1 lb. of water substance at absolute pressures of 100, 80, 60 and 40 lb. per sq. in., and from the table plot the water and dry steam lines on the entropy-temperature diagram.

Note how nearly these lines are straight lines, and remain straight up to large pressures.

1 lb. of dry steam at 100 lb. pressure is condensed at constant volume to 40 lb. Draw the constant volume line upon the diagram and determine the dryness and entropy of the substance at the above pressures.

4. A 500 kilowatt steam turbine using dry steam at a pressure of 140 lb. per sq. in. consumes 22·6 lb. of steam per kilowatt hour

and condenses at a pressure of 1 lb. per sq. in. The condensing water measures 4960 lb. per min. and its rise of temperature is 18·3° C. Determine the dryness of the steam as it leaves the turbine and find the loss of work in the cylinder of the turbine due to the expansion not following the adiabatic line on the entropy-temperature diagram.

5. 1 lb. of wet steam occupies a volume of 1 cu. ft. at a pressure of 100 lb. per sq. in. It is allowed to expand, its pressure being kept constant, until its volume is 4 cu. ft., and at this volume its pressure is allowed to fall to 50 lb. per sq. in. It is condensed at this pressure until its volume is again 1 cu. ft. and its pressure is then raised, at this volume, to its original value. Draw the p-V and the ϕ-T diagrams for the cycle, and calculate the changes in E and in I and the amounts of heat added or removed, for each stage of the cycle.

6. 1 lb. of steam at a pressure of 100 lb. per sq. in. and dryness fraction 0·75 is enclosed in a cylinder by a frictionless piston. It may expand in three ways: (1) according to the law $pV = \text{const.}$, (2) adiabatically, (3) with constant dryness fraction. If expansion takes place until the pressure is 50 lb. per sq. in., find, for each case, the external work done, the heat interchange during the expansion and the final dryness fraction.

7. The diagram is taken from a double acting engine running at 180 R.P.M. and having a stroke volume of 5·5 cu. ft. and clearance volume 0·5 cu. ft.

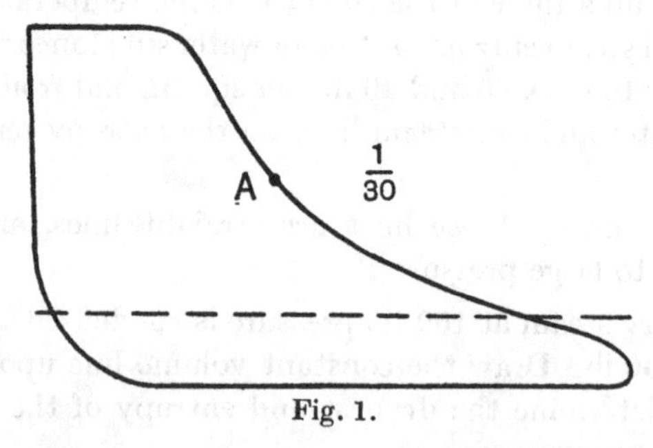

Fig. 1.

The air pump discharge of the engine is 66 lb. per min. Estimate the dryness of the steam at the point A. Atmospheric pressure 15 lb. per sq. in.

8. Draw the entropy-temperature diagram for steam, between the temperatures 60° C. and 160° C., and on it shew the lines of constant volume for $V=20$, 30, and 40 cu. ft. respectively, the lines of constant total energy for $I=400$, 500, and 600 thermal units respectively, and the lines of constant dryness for $q=0{\cdot}6$ and 0·8 respectively.

For each curve three points, if properly selected, are sufficient to give the general direction.

PAPER VIII

1. A pump is placed above a tank containing water at a temperature of 90° C. The weight of the suction valve is 2 lb. and its diameter $1\frac{1}{2}$ in. Find the maximum height above the tank at which the pump may be placed so that it will draw water; the barometer stands at 30 in. and the pump may be assumed perfect and without clearance.

2. 1 lb. of dry steam at a pressure of 120 lb. per sq. in. is allowed to condense at constant volume until its pressure falls to 20 lb. per sq. in. What is then its dryness fraction, and how much heat has been rejected?

What is the final state of the steam if it is allowed to expand adiabatically from its given initial pressure to a pressure of 20 lb. per sq. in.?

3. A closed vessel of 50 cu. ft. capacity contains 750 lb. of water and steam at 100° C., and there is no air present. The temperature is raised by admitting, through a stop valve, superheated steam from a supply at 150 lb. per sq. in. and 250° C. Assuming that there is no loss of heat by conduction, find the weight of steam required to bring the temperature to 150° C.

(Take the specific volume of water at these temperatures to be 0·017 cu. ft. per lb.)

4. An engine exhausts at 120° C., the atmospheric pressure being normal: at the moment of release the wetness is 0·85. Find the thermal units per lb. lost owing to incomplete expansion, assuming the constant volume lines straight over the range considered.

5. With the help of Mollier's diagram, or otherwise, solve the following:

(i) Superheated steam at a pressure of 100 lb. per sq. in. and temperature 200° C. expands in the H.P. cylinder of a compound engine and on leaving the cylinder is throttled before entering the L.P. cylinder. The engine works on the Rankine cycle and uses 11 lb. of steam per I.H.P. hour. Assuming that the same amount of work is done in both cylinders and that the lowest pressure is 2 lb. per sq. in., determine the range of pressure in each cylinder, and the fall of pressure and change of entropy in throttling.

(ii) Assuming frictionless adiabatic expansion in a properly designed nozzle, find the area of the opening of the nozzles for a 20 H.P. turbine using 15 lb. of steam per H.P. hour, two nozzles being fitted. The steam is initially dry and expands from a pressure of 100 lb. per sq. in. to a pressure of 2 lb. per sq. in.

6. *A* and *B* are two points on the expansion line of an indicator diagram, taken from a steam-engine cylinder. At *A* the volume is 0·4 cu. ft., the pressure is 65 lb. per sq. in. and the dryness of the steam is 0·75. At *B* the volume is 0·9 cu. ft. and the pressure is 37 lb. per sq. in. The volumes given include clearance. Find how much steam is present in the cylinder and its dryness at the point *B*.

Also, assuming that the curve of expansion is of the form pV^n = a constant, find the value of n, and determine how much heat has been received by the steam between the points *A* and *B*.

7. In a Stirling engine fitted with a perfect regenerator, the maximum pressure is 115 lb. per sq. in. and the minimum 15 lb. per sq. in., the higher and lower temperatures being 400° C. and 20° C. respectively. A perfectly reversible steam engine uses dry steam between the same limits of pressure. Compare the efficiency of the two engines; and if the piston speed and stroke be the same in each, compare the piston areas for equal power.

8. A vertical non-conducting cylinder, with a cross-sectional area of 1 sq. ft., contains 1 lb. of water at a temperature of 15° C. A close-fitting piston rests on the water, its weight being such that the water is under a pressure of 100 lb. per sq. in. An

electric heater placed in the bottom of the cylinder communicates heat to the water at the rate of 50 watts. In what time will the piston rise through a distance of 2 ft.?

If the piston had been originally fixed at a height of 2 ft. above the surface of the water, how long would it take to fill the cylinder with steam mixture at 100 lb. per sq. in.?

PAPER IX

1. Use Mollier's ϕ-I diagram to solve the following cases:

(i) Steam of dryness 0·85 expands adiabatically and reversibly from a pressure of 200 lb. per sq. in. to a pressure of 2 lb. per sq. in. Find the heat drop and the final dryness.

(ii) Steam expands at constant dryness, 0·85, between the above limits of pressure. Find the heat drop and increase of entropy.

(iii) Dry steam is throttled from the 200 lb. per sq. in. pressure to 2 lb. per sq. in. Find the change in entropy and the final state of the steam.

2. Calculate (i) the loss of entropy in 10 lb. of steam of pressure 150 lb. per sq. in. of dryness 0·95 when condensed to water at 20° C., (ii) the gain in entropy of 1 lb. of air when suddenly expanded to three times its original volume without doing work.

3. By means of Mollier's diagram, or otherwise, solve the following cases:

(i) Steam is formed at a pressure of 225 lb. per sq. in. and is 90 per cent. dry. If the lowest pressure available is 2 lb. per sq. in., compare the work done per lb. of steam in the two cases: (*a*) steam admitted to the cylinder as formed, (*b*) steam throttled adiabatically to 150 lb. per sq. in. and then admitted to the cylinder.

(ii) Superheated steam of pressure 150 lb. per sq. in. and temperature 250° C. expands in the H.P. cylinder of a compound engine, and on leaving is throttled before entering the L.P. cylinder. The lowest pressure is 2 lb. per sq. in.

and the amount of work done in both cylinders is the same. If the engine works upon the Rankine cycle and consumes 10 lb. of steam per I.H.P. hr., determine the range of pressure in each cylinder and the fall of pressure and increase of entropy in throttling between the cylinders.

4. Calculate the change of entropy when:

(i) 10 lb. of air is compressed to one-third of its volume according to the law $pV^{1\cdot3}=\text{const.}$

(ii) 1 lb. of steam of dryness 0·9 at a pressure of 120 lb. per sq. in. is throttled down to a pressure of 10 lb. per sq. in.

5. By means of Mollier's diagram, or otherwise, solve the following cases:

(i) Steam expands adiabatically and reversibly from a pressure of 150 lb. per sq. in. and temperature 300° C. to a pressure of 2 lb. per sq. in.: find the dryness at this pressure and the heat drop per lb. of steam.

(ii) Steam expands from 200 lb. per sq. in. to 2 lb. per sq. in. at constant dryness 0·9: find the heat drop and change of entropy per lb. of steam.

(iii) Dry steam is throttled from 200 lb. per sq. in. to 50 lb. per sq. in.: find the state of the steam and the change of entropy per lb.

6. The working substance of an engine consists of 1 lb. of water and 1 lb. of air, the water and air being separated by a light frictionless piston fitted in the cylinder. The cycle starts with an isothermal expansion at a temperature of 150° C. from a pressure of 100 lb. per sq. in. and when all the water has become steam the pressure is reduced at constant volume to 10 lb. per sq. in. From this point the cycle is completed by a constant pressure line and a constant volume line. Determine the work done in the complete cycle and the efficiency of the cycle.

7. A boiler of 1200 cu. ft. capacity contains 40,000 lb. of steam and water at a pressure of 50 lb. per sq. in.: if the efficiency of the boiler is 65 per cent., calculate the weight of coal to be burnt to raise the pressure to 150 lb. per sq. in. The calorific value of the fuel is 7500 TH.U.

8. By means of Mollier's ϕ-I diagram, or otherwise, solve the following cases:

(i) 1 lb. of dry steam at a pressure of 200 lb. per sq. in. is allowed an adiabatic total heat drop of 150 TH.U., divided into three equal stages. Find the pressure and dryness at the end of each stage.

(ii) Steam expands at constant dryness 0·9 between the pressures of 150 and 2 lb. per sq. in. Find the heat drop and the increase of entropy.

(iii) Steam at a pressure of 150 lb. per sq. in. is throttled until at the pressure of 10 lb. per sq. in. the temperature is 100° C. Find the original dryness of the steam and the change of entropy due to throttling.

PAPER X

1. Find the dryness fraction and the volume of 1 lb. of steam at a pressure of 100 lb. per sq. in., when its internal energy is 380 TH.U.

Draw a curve shewing the relation between pressure and volume of this pound of steam as it expands to a pressure of 20 lb. per sq. in., under conditions which maintain its internal energy at the given value.

2. A non-conducting cylinder whose cubic content is 4 cu. ft. is divided by a rigid non-conducting partition into two equal volumes. Each half of the cylinder contains 1 lb. of steam, in the one end the pressure is 150 lb. per sq. in., in the other end it is 50 lb. per sq. in. The partition collapses, allowing free mixing of the steam. Find its ultimate pressure and condition.

Find also the total change of the entropy of the contents of the cylinder.

3. Steam at a pressure of 100 lb. per sq. in. and at a temperature of 400° C. is blown from a supply pipe into a boiler of capacity 500 cu. ft. containing 200 lb. of water and steam at a pressure of 20 lb. per sq. in. Determine the weight of steam which must be blown in in order to bring the pressure up to 60 lb. per sq. in.

The volume of water may be neglected.

4. In a steam-jacketed cylinder heat is supplied from the jacket at such a rate as just to keep the cylinder steam dry and saturated during expansion. Assuming that the latent heat of steam at t° C. is equal to $606-0{\cdot}7t$ TH.U., find how much heat is supplied by the jacket steam per pound of cylinder steam when the latter expands in this way from a pressure of 180 lb. to a pressure of 40 lb. per sq. in.

Find also the change of internal energy in the cylinder steam and deduce the external work done by it. Compare this with the value derived from the area of the expansion curve $p.V^n=$a constant, where n has the value 1·06.

5. Heat passes from 1 lb. of water at 100° C. to 1 lb. of water at 20° C., so that ultimately their temperatures are equalised. Calculate the final temperature and the change of entropy (if any) (1) when the passage of heat takes place by direct conduction, (2) when the maximum amount of work theoretically available is developed. The specific heat of water may be taken as constant and equal to unity.

6. A boiler evaporates water to steam of dryness 0·95 at a pressure of 150 lb. per sq. in. Determine the efficiency of the boiler if 20,000 lb. of water are evaporated per ton of coal from feed water at 80° C. The calorific value of the coal is 8000 TH.U.

The stop-valve of a boiler containing 20,000 lb. of water and 150 lb. of steam at a pressure of 150 lb. per sq. in. is closed and the boiler allowed to cool. Shew that when the pressure reaches 50 lb. per sq. in. nearly 97 lb. of steam have been condensed.

7. Two turbines are fed from the same boiler which supplies steam at 150 lb. pressure and of dryness 96 per cent. Turbine *A* takes steam direct from the boiler, and turbine *B* takes its supply through a reducing valve, the steam-chest pressure being 120 lb. per sq. in. Assuming adiabatic expansion in both turbines, find how much circulating water each will require, per lb. of steam used, if the pressure in the condenser is 1·5 lb. per sq. in., the inlet and outlet temperatures of the circulating water 15° C. and 43° C. respectively and the hot-well temperature 45° C.

8. Describe the action of thermal accumulators for the storage of the heat of exhaust steam.

An exhaust steam turbine requires 3000 lb. of steam per hour, and is to be kept working for 10 min. by steam supplied from a thermal accumulator. Determine the capacity of the accumulator if its pressure is allowed to fall from 2 lb. per sq. in. above to 1 lb. below atmosphere (15 lb. per sq. in.): assume the lagging is perfect.

PAPER XI

1. An ejector pumps water from the bilges of a ship, discharging 20 tons of water per hour at a mean height of 12 ft. above the level in the bilges. The boiler pressure is 200 lb. per sq. in., the temperature of discharge is 32° C. and that of the bilge 10° C. The temperature of the feed water to the boiler is 15° C. Find the mass of steam used per ton of bilge water discharged and the efficiency of the ejector.

2. The temperature inside a surface condenser is 46·6° C. and the pressure 2 lb. per sq. in. The steam space is of 40 cu. ft. capacity. Shew that the weight of air present is 0·093 lb. and is 53 per cent. of the weight of steam present.

An engine with a condenser is of 50 I.H.P. and consumes 16 lb. of steam per I.H.P. hour. Assuming that the exhaust steam is of dryness 0·7, and enters the condenser free of air, determine the proper volume of the air pump cylinder making 120 strokes per min. and the rate of leakage of air into the condenser when the temperature of the condenser is 50° C. and the pressure 1·9 lb. per sq. in.

3. The working substance of an engine receives 200 TH.U. as its temperature rises from 50° to 250° C. at a rate proportional to the rise of temperature; it then receives 500 TH.U. at a constant temperature of 250° C. and afterwards 100 TH.U. while the temperature is falling from 250° C. to 50° C., the rate of supply being proportional to the change in temperature. The heat that is not utilised is all rejected at the temperature of 50° C. Sketch the entropy-temperature diagram for the cycle and give values of the changes of entropy. Determine the efficiency of the cycle.

4. In a single-acting engine working without compression the volume swept out by the piston is $\frac{1}{4}$ cu. ft., the clearance space is $\frac{1}{10}$ cu. ft., and cut-off is at $\frac{3}{10}$ of the stroke. The engine exhausts into the atmosphere and takes 0·05 lb. of dry saturated steam at 120 lb. per sq. in. pressure per revolution. Find the heat given to the walls of the cylinder by condensation of steam during admission.

5. The output of an engine is varied by altering the ratio of expansion while the speed and boiler pressure are kept constant and the consumption is found to vary with the B.H.P. as in the following table:

B.H.P.	20	30	40	50	60
Consumption in lb. of water	970	1150	1400	1740	2380

Shew that the highest thermal efficiency is obtained when the cut-off is arranged to give approximately 46 B.H.P.

6. A boiler contains 2 tons of water and steam under a pressure of 125 lb. per sq. in., the water volume being half the steam volume. Assuming that there is no transfer of heat to or from the boiler, find how much steam must be blown off in order to reduce the pressure to 100 lb. per sq. in. What fraction of the whole volume will the water occupy at the reduced pressure?

7. 1 lb. of steam at a pressure of 200 lb. per sq. in. is superheated to a temperature of 400° C., and is then expanded adiabatically until its volume is 45 cu. ft. Condensation then takes place at constant volume until its temperature is 50°, after which condensation takes place at constant temperature. Assuming that the specific heat of superheated steam at constant pressure is equal to 0·48 and that the specific heat of water is constant and equal to unity, sketch the temperature-entropy diagram, and find (1) the temperature at release, (2) the work done, (3) the efficiency.

8. Two boilers, each with 500 cu. ft. of water and steam space, are under steam at pressures of 200 and 100 lb. per sq. in. respectively. They are both one-half full of water. Communication is opened between the boilers. Find the total quantity of heat that must be given to or rejected from the two boilers to bring the common pressure to 180 lb. per sq. in. What is the weight of water now present?

PAPER XII

1. In an ideal steam power plant, using saturated steam and isentropic expansion between the temperatures T_1 and T_2, progressive bleeding is applied so that the feed water is heated to the temperature T_1. Shew that at any intermediate temperature T the ratio of the amount of feed water being heated to the amount of feed water at T_1 is $\frac{\phi-\phi_{\omega 1}}{\phi-\phi_{\omega}}$, where ϕ, ϕ_{ω} and $\phi_{\omega 1}$ refer to the entropy of the expanding steam, the feed water at T and the feed water at T_1 respectively.

If the expansion is from a temperature of 200° C. and 80 per cent. dryness to a temperature of 40° C., determine the proportion of steam progressively bled over the whole range.

2. Steam expands isentropically from a pressure of 200 lb. per sq. in. and a temperature of 400° C. to such a temperature that its volume per lb. is 160 cu. ft. Determine graphically, or otherwise, the temperature, dryness and I of the steam at the end of the expansion.

3. In a steam power plant the steam is caused to expand isothermally from a temperature of 350° C. and a pressure of 200 lb. per sq. in. to a pressure of 25 lb. per sq. in. in the high pressure cylinder and thence isentropically to 0·8 lb. per sq. in. in the low pressure cylinder at which pressure it passes to the condenser.

Determine the work done in each cylinder per lb. of steam and the efficiency of the plant. How much heat is received per lb. of steam during the isothermal expansion?

4. A steam engine takes in 0·180 lb. of steam per stroke. The stroke volume is 2 cu. ft. and the clearance volume is 0·2 cu. ft. Compression begins at 30 per cent. from the end of the exhaust stroke, the pressure then being 15 lb. per sq. in. and the steam dry. Cut-off takes place at 11 per cent. and release at 85 per cent. of the working stroke, the corresponding steam pressure being 140 and 35 lb. per sq. in. Find the total weight of steam present during expansion, the dryness at cut-off and

release. Assuming pV^n = constant during expansion, find for the expansion period the work done and the interchange of heat between the steam and cylinder.

5. The table gives the pressure and volume per lb. of steam during its expansion in the cylinder of a steam engine:

85	65	45	25	16 lb. per sq. in.
4·00	5·23	7·56	13·60	21·25 cu. ft.

Plot to scale this expansion on a T-ϕ diagram and hence estimate the heat received from the cylinder walls during the expansion.

6. For engines which are required to be working intermittently heat energy is accumulated in the following manner: large cylinders in communication with the boiler are completely filled with water, and when the engines are at rest the heat developed is employed in raising the temperature of the water, and the stored heat is ready to produce steam when a diminution of the pressure is permitted to take place. The highest pressure to which the cylinders are to be exposed is 150 lb. per sq. in. and the pressure is worked down to the lower limit of 120 lb. per sq. in.: how many cubic feet of water will be required in the accumulators per horse-power-hour, assuming the average consumption of steam to be 18 lb. per horse-power-hour?

7. A boiler of 200 cu. ft. capacity contains equal volumes of water and steam when the pressure is 30 lb. per sq. in. Find the quantity of heat which must be given to the boiler to raise the pressure to 120 lb. per sq. in. If the efficiency of the boiler is 69 per cent. and the calorific value of the coal 8500 TH.U. per lb., how much coal will have to be consumed?

8. Two boilers, whose volumes are in the ratio of 2 to 3, are under steam at pressures of 200 and 150 lb. per sq. in. respectively. The first boiler is $\frac{1}{4}$ full of water, and the second is $\frac{1}{2}$ full. A connection is opened between them, and the resulting pressure is 180 lb. per sq. in. abs. Assuming that no steam has escaped, find the total quantity of heat, per unit of total volume, rejected from, or supplied to, the whole system during the equalisation of the pressures.

PAPER XIII

1. The boiler-room of a battleship contains six Yarrow large-tube boilers, the dimensions of each of the grates being 7 ft. × 8 ft. 6 in. 40 lb. of coal are being burnt per sq. ft. of grate per hour, the composition of the coal being C, 90 per cent.; H, 4 per cent.; O, 2 per cent. The fan for the forced draught produces a pressure in the suction duct of 0·04 in. of water below atmospheric pressure. If the supply of air by the fan is 100 per cent. in excess of the minimum required for complete combustion, calculate the section of the air ducts, the temperature of the air being 15° C., the specific volume of the air at that temperature and atmospheric pressure being 13 cu. ft. per lb.

2. In the case of a liquid and its vapour expanding through a throttle valve from one region to another in which the pressures are kept constant, if there is only liquid present on the high pressure side, the pressure and temperature corresponding to saturation, and if the specific heat of the liquid is S, shew that the gain of entropy per unit mass will be

$$S\left\{\frac{T_1-T_2}{T_2}-\log\frac{T_1}{T_2}\right\},$$

where T_1 and T_2 are the absolute temperatures before and after passing the valve, and the specific volume of the liquid may be neglected.

3. Steam at a pressure of 200 lb. per sq. in. and temperature 300° C. expands in an engine to a pressure of 50 lb. per sq. in. and is then taken and reheated at this pressure to its original temperature of 300° C. It is then expanded further in the engine down to the condenser pressure at 1 lb. per sq. in. Shew that the introduction of the reheating stage increases the efficiency and determine by what percentage.

The expansion of the steam in the engine may be assumed adiabatic and of constant entropy.

4. The motion of a piston in a cylinder is controlled by a spring which, together with the atmosphere, exerts a pressure of 50 lb. per sq. in. on the piston when it is at the end of its stroke

and 100 lb. per sq. in. when the swept volume is 1 cu. ft. Connected to the end of the cylinder is a boiler which supplies steam through a stop valve at 100 lb. per sq. in. and dryness 0·8. The clearance space between the stop valve and the piston is 0·25 cu. ft. and initially this is filled with dry saturated steam at 20 lb. per sq. in.

If the stop valve is opened, find the steam supplied before the piston moves, and also the total amount of steam supplied.

Conductivity of the cylinder and piston, and the weight and inertia of the piston and spring are to be ignored.

5. In a turbine consisting of a high pressure and a low pressure section, steam is supplied at 225 lb. per sq. in., superheated to 350° C. and expands in the H.P. section to 25 lb. per sq. in.: the steam is then reheated at constant pressure to 250° C. and expanded in the L.P. section to 0·4 lb. per sq. in. Find the efficiency of the cycle, assuming adiabatic expansion.

In an alternative scheme the steam is initially superheated to 450° C. but its temperature is reduced to 350° C. before entering the turbine by passing through a heat interchanger through which the steam also passes between the H.P. and the L.P. sections. Assuming that the interchanger is perfect and the pressures at the various stages are the same in the two cycles, compare the efficiencies obtained by the two schemes.

6. Shew that the gain ρ of internal energy which takes place when 1 lb. of water is just completely evaporated at constant pressure p (saturation temperature $T°$ abs.) is given by

$$\rho = L\left(1 - \frac{p}{T}\frac{dT}{dp}\right).$$

Using the steam tables for L and the connection between p and T, find the value of ρ at $T = 373°$ C. abs. (100° C.).

7. An injector delivers 1000 gallons of water per hour from a supply tank where the temperature is 23° C. to a boiler in which the pressure is 100 lb. per sq. in. The injector is supplied with steam at boiler pressure and dryness 0·9, and the pressure at the outlet of the steam supply nozzle is 0·6 of the boiler

pressure. The temperature of the boiler feed is 80° C. The supply tank, the injector and the boiler may be assumed to be at the same level.

Find the weight of steam used in the injector per hour, and the area of the steam orifice, and prove that the injector will work against the given boiler pressure.

State carefully what assumptions you have made in dealing with the problem.

8. So long as the weight of bled steam is not greater than that necessary to raise the temperature of the feed water to that of the bled steam, shew that bleeding must increase the efficiency of a steam plant whatever weight of steam is bled and at whatever temperature, intermediate between supply and exhaust.

Determine the increase in theoretical efficiency when the best proportion of steam is bled at 30 lb. per sq. in. in a turbine plant working between 200 lb. per sq. in. 300° C. and 1 lb. per sq. in.

PAPER XIV

1. An engine develops 300 horse-power on a consumption of 65 lb. of steam per min. The steam is superheated to a temperature of 250° C. at a pressure of 120 lb. per sq. in. The exhaust pressure is 2 lb. per sq. in. Find the thermal efficiency of the engine and determine how many pounds of steam the engine would use per minute if working upon the Rankine cycle between the same limits of temperature and pressure.

2. A steam engine is supplied with steam at 120 lb. per sq. in. pressure and uses 15·7 lb. per I.H.P. hour when the condenser pressure is 2 lb. per sq. in. Compare the thermal efficiency of the engine with the efficiency of a Rankine cycle with the same range of pressures.

3. An engine is supplied with dry saturated steam at a pressure of 130 lb. per sq. in., and cut-off occurs at 5·1 per cent. of the stroke. Determine the work done per lb. of steam and the efficiency if the engine works on the theoretical (Rankine) cycle.

(Exhaust temperature is between 65° and 75° C.)

4. A double-acting compound steam engine running at 121 R.P.M. has piston areas of 50 and 160 sq. in. and a stroke of 18 in. In a test the mean effective pressures were 48 and 15 lb. per sq. in. respectively. The engine used 1000 lb. of dry saturated steam per hour at a pressure of 170 lb. per sq. in. The air pump discharge temperature was 45° C. and 400 lb. of circulating water per minute was raised 22° C.

Draw up a heat balance sheet and determine the thermal efficiency of the engine, and the Rankine cycle efficiency for the same limits of temperature.

5. An engine takes dry steam at 160 lb. per sq. in. and its exhaust pressure is 4 lb. per sq. in. The consumption is 13·5 lb. of steam per I.H.P. hour. What is the thermal efficiency?

How much steam is required per I.H.P. hour by an engine working on the Rankine cycle between the same limits of pressure, and what is the thermal efficiency of this engine?

6. 1 lb. of steam is initially at 200 lb. per sq. in. pressure and its dryness fraction is 0·9. It expands adiabatically until its pressure is 100 lb. per sq. in. and is then condensed at constant volume until its pressure falls to 50 lb. per sq. in. The steam is then compressed adiabatically until its pressure is again 200 lb. per sq. in., and finally is evaporated at this pressure until it reaches its initial volume. Draw the ϕ-T and the p-V diagrams for the cycle, and tabulate, for each stage, the heat supplied and the work done.

7. 10 lb. of steam of dryness 0·96 and at a pressure of 120 lb. per sq. in. expands adiabatically to a pressure of 12 lb. per sq. in. and is then cooled at constant volume to a pressure of 2 lb. per sq. in.: determine the dryness of the steam at this pressure. If the steam is now condensed at constant pressure until it reaches its original volume and then heated up at constant volume to its original pressure, determine the work done in the cycle, the heat taken in, and the efficiency.

Why is this efficiency lower than the highest efficiency possible with the given temperatures?

8. Dry steam is admitted to an engine cylinder at a pressure of 100 lb. per sq. in. and 20 per cent. of it is condensed during admission, without fall of pressure. The steam expands down to a pressure of 20 lb. per sq. in. and during expansion one half of the heat absorbed by the walls is returned to the steam at a uniform rate as the temperature falls. Find the dryness fraction of the steam at the end of expansion. Find also the amount of work obtainable in the cylinder per pound of steam from the returned heat.

Shew that if the steam expand adiabatically to the same final volume the final pressure will be a little below 18 lb. per sq. in.

PAPER XV

1. Dry saturated steam at 150 lb. per sq. in. enters the H.P. cylinder of a compound engine and expands down to 50 lb. per sq. in., when it exhausts to a large receiver at 30 lb. per sq. in. It is now admitted to the L.P. cylinder, expands adiabatically down to 10 lb. per sq. in., and is exhausted at 2 lb. per sq. in. Neglecting heat losses, estimate the work done per lb. of steam and compare it with the Rankine cycle between the upper and lower limits of pressure, i.e. 150 and 2 lb. per sq. in.

2. In a compound engine the admission pressure in the high pressure cylinder is 150 lb. per sq. in., the release pressure 65 lb., the receiver pressure 30 lb., the release pressure in the low pressure cylinder 8 lb. and the back pressure 2 lb. Assuming the walls non-conductors of heat, the receiver of large volume, the boiler steam dry, neglecting the effects of clearance, etc., and taking account solely of 'drop', estimate the total work done, and compare it with that which would be done if there was no drop in either cylinder.

3. Assuming that the steam expands hyperbolically in the cylinder, estimate the I.H.P. of an engine running at 150 R.P.M. and receiving steam at 90 lb. per sq. in. pressure and exhausting at 4 lb. per sq. in. with a ratio of expansion of 10 at the back end of the cylinder and 8 at the front end. The diameters of the piston and piston rod are 20 in. and $2\frac{1}{2}$ in. respectively and the length

of stroke is 24 in. What is the thermal efficiency of the engine, if supplied with 15 lb. of steam per I.H.P. per hr., of dryness 0·85, from feed water at 35° C.?

4. Two trials of an engine are made. In both the steam pressure is 195 lb. per sq. in. and the exhaust pressure is 2·3 lb. per sq. in. In the first trial the steam is dry and saturated and the consumption is 14·1 lb. per I.H.P. per hr. In the second trial the steam is superheated and its temperature is 230° C., and the consumption is 12·8 lb. per I.H.P. per hr. Find in each case the heat used per I.H.P. per min. Find also, for the corresponding Rankine cycle in each case, the heat used per I.H.P. per min. and the steam used per I.H.P. per hr.

5. In a 20 min. test a double-acting single-cylinder engine makes 3920 revolutions and uses 410 lb. of steam. The cylinder diameter is $7\frac{1}{4}$ in., the stroke is 12 in. and the clearance volume is 10 per cent. of the volume swept through by the piston. Steam is supplied at 160 lb. per sq. in. and the cushion steam is estimated to be just sufficient to fill the clearance volume when dry and saturated at 80 lb. per sq. in. The ratio of expansion, measured on the diagram with no allowance for clearance, is 2·54. Find the actual ratio of expansion, the weight of steam present during expansion, and its dryness at the point of cut-off.

At points on the diagram corresponding to 7, 9 and 11 in. of stroke the pressure is 112, 81·3 and 69 lb. per sq. in. respectively; find the state of the steam at these points and plot the expansion line on the ϕ-T diagram.

6. An engine is required to work at the rate of 500 I.H.P. with a piston speed of 500 ft. per min. Assuming that cut-off is to take place at 25 per cent. of the stroke, that the expansion curve is a rectangular hyperbola, that the initial pressure is 140 lb. per sq. in. and that the back pressure is 5 lb. per sq. in., calculate the size of the cylinder, and the number of revolutions per min., the stroke being taken equal to twice the diameter. You may assume that there is no compression, and may neglect the effect of clearance. Diagram factor 0·9.

7. A boiler delivers to a steam-pipe dry steam at 160 lb. per sq. in. pressure. There is a separator and steam trap at the engine

end of the pipe, and the steam is supplied to the engine, dry and at 145 lb. pressure, while for each 1 lb. of steam supplied by the boiler 0·05 lb. of water is drawn off at the trap. This water is returned to the boiler at the temperature at which it leaves the pipe. Assuming that the steam is used as in a Rankine cycle and that the pressure of the exhaust is 2·5 lb. per sq. in., find the percentage reduction of output per lb. of coal caused by the pipe.

8. Steam is expanded in an engine from a supply pressure of 100 lb. per sq. in. to 10 lb. per sq. in. It is then condensed at 2 lb. per sq. in. During admission to the cylinder there is an immediate condensation on the walls of 20 per cent. of the steam supply, but during expansion half of the heat given to the walls is yielded up again to the steam, the heat yielded up per degree fall of temperature being constant.

Draw the entropy-temperature diagrams for the steam during expansion and condensation, and calculate the dryness of the steam at the end of the expansion. The diagram need not be drawn strictly to scale, but the general shape of the lines should be correctly shewn, and entropy and temperature values marked at important points.

PAPER XVI

1. A single-acting engine, working without compression and exhausting into the atmosphere, at a pressure of 15 lb. per sq. in. is supplied with 0·08 lb. of dry saturated steam per revolution. The steam pressure is 150 lb. per sq. in. The cylinder volume is 0·4 cu. ft., the clearance volume is 0·1 cu. ft. and cut-off occurs at 0·2 of the stroke. The clearance steam is dry. Find how much steam is condensed during admission.

2. The initial pressure of steam supplied to a double-acting condensing engine is 100 lb. per sq. in. and cut-off is at $\frac{3}{10}$ of the stroke. The condenser pressure is 2 lb. per sq. in. Assuming that the expansion line has the form $pV = \text{constant}$, shew that the mean pressure in the cylinder is 64 lb. per sq. in.

Assuming the area of the actual diagram to be 0·85 of the trial diagram and allowing a piston speed of 600 ft. per min., determine the diameter of cylinder to develop 350 H.P.

3. In the high pressure cylinder of a compound engine of which the hot-well discharge per minute was 29 lb. and the revolutions per min. 24, the absolute pressures in lb. per sq. in., at the beginning of compression, at cut-off and at release were 14·8, 64 and 15·2 respectively, and the volumes, including clearance, shewn by the indicator card were 1·52, 2·92 and 13·24 cu. ft. Assuming the steam dry at the point of compression, estimate the dryness fractions at the points of cut-off and release.

4. A Rankine engine takes steam at a pressure of 200 lb. per sq. in., and exhausts at a pressure of 2 lb. per sq. in. Calculate the thermal efficiency and the amount of steam used per I.H.P. hour (1) when the steam is dry and saturated, (2) when it is superheated to 295° C. Find also the efficiency of the corresponding Carnot engine in each case.

5. An engine is supplied with dry steam at a pressure of 180 lb. per sq. in. and the exhaust pressure is 4 lb. per sq. in. The consumption is 14 lb. per I.H.P. per hour. Find the thermal efficiency.

Find also the amount of steam required per I.H.P. per hr. by a Rankine engine working between the same limits of pressure, and its thermal efficiency.

What amount of energy is supplied per I.H.P. per min. in each case?

6. A boiler supplies steam at a pressure of 200 lb. per sq. in., the dryness fraction being 0·9. After passing through a reducing valve, with a reduction of pressure to 160 lb. per sq. in., the steam enters the H.P. cylinder of a compound engine where it expands adiabatically to 40 lb. per sq. in. It is exhausted from this cylinder at 22 lb. per sq. in. entering the L.P. cylinder, where it expands adiabatically to 8 lb. per sq. in., exhaust taking place at 2 lb. per sq. in. Estimate the volume of the steam per lb. at cut-off in each cylinder and at the beginning of exhaust in the L.P. cylinder.

7. A steam engine, working on the Rankine cycle, uses steam (dry saturated) at 175 lb. per sq. in., the temperature of the condenser being 60° C. Find the work (in I.H.P. hours) per lb. of steam, and the thermal efficiency.

If the exhaust steam be utilised to maintain on the Rankine cycle a SO_2 engine, the steam engine condenser acting as the boiler, find the additional work (in I.H.P. hours) obtained per lb. of steam, and the total thermal efficiency of the combined plant, the SO_2 condenser being at 20° C. Give also the ratio of pounds of SO_2 used to pounds of steam. The SO_2 supply may be taken as dry saturated.

(N.B. For SO_2 at 60° C., $I_l = 21{\cdot}50$, $I_v = 85{\cdot}10$, $\phi_v = 0{\cdot}2617$. Use table for SO_2 at 20° C.)

8. An engine uses two working fluids, steam and aether vapour. The water is evaporated to dry steam at 50 lb. per sq. in. and after expanding in the engine is released at 7 lb. per sq. in. and condensed at that pressure giving up its heat to the aether. The aether vapour is formed at temperature 80° C. and after expanding doing work, is itself condensed at the temperature of 50° C.

Determine the work done in the engine by the aether and compare it with the work that would be got from the steam, if it had exhausted at the pressure 2 lb. per sq. in. of the condenser instead of at the pressure 7 lb. per sq. in. of release.

Table of properties of aether:

Temp. °C.	Pressure lb. per sq. in.	Latent heat of 1 lb. of vapour	Volume of 1 lb. of vapour	ϕ_l	ϕ_v
80	58·5	157	1·55	0·133	0·578
50	17·6	159	3·39	0·086	0·577

Specific heat of liquid aether 0·513.

PAPER XVII

1. A steam engine works upon a cycle similar to the Rankine with the exception that the adiabatic expansion is replaced by a stage in which the steam as it expands receives, per unit fall of temperature, heat proportional at any instant to the absolute temperature. If the whole quantity of heat received in this way is Q, draw the ϕ-T diagram and find the efficiency of the cycle.

Shew that, if the steam reaches T_2 dry,

$$Q = \frac{T_1 + T_2}{2} (\phi_{2s} - \phi_{1s}).$$

If the rate at which heat is received $=T^2-aT$, find the heat received and rejected and the value of a $(a<T_1)$ if the steam at the end of expansion is just dry. Find the efficiency of this cycle.

2. Shew that the area enclosed by a complete reversible cycle on the entropy-temperature diagram is equal to the area enclosed by the corresponding p-v diagram.

Dry steam issues from a boiler at 150 lb. per sq. in., and during admission to the engine cylinder 10 per cent. is condensed. During expansion one-half the heat abstracted during admission is returned at a uniform rate as the temperature falls. The pressure at the end of expansion is 10 lb. per sq. in. If the fall of pressure to the condenser at 2 lb. per sq. in. may be taken to be at constant volume, sketch the entropy-temperature diagram.

Find the percentage of water present at the end of the stroke and also just before condensation in the condenser.

3. In a compound engine the admission pressure in the high pressure cylinder is 115 lb. per sq. in., the release pressure 50, the receiver pressure 25, the release pressure in the low pressure cylinder 6 and the back pressure 2 lb. per sq. in. Assuming the walls non-conductors of heat, the receiver of large volume, the boiler steam dry, neglecting the effects of clearance, etc., and taking account solely of 'drop', estimate the total work done, and compare it with that which would be done if there was no drop in either cylinder.

4. A steam engine working on the Rankine cycle uses dry steam at a pressure of 150 lb. per sq. in. and the temperature of the condenser is 50° C. Calculate the work done per lb. of steam, and the thermal efficiency of the engine.

If the engine is steam-jacketed, estimate per lb. of feed steam the heat required from the jackets to keep the steam dry during expansion, the work done and the efficiency.

5. Dry steam is admitted to a cylinder at a pressure of 80 lb. per sq. in. and the condensation during admission is 20 per cent. of the whole steam supply. During expansion heat is reabsorbed at such a rate that a constant number of units of heat are taken in for each degree drop of temperature. The final pressure is 2 lb. per sq. in. and the final dryness of the steam is 0·85: find the amount of heat reabsorbed.

6. Distinguish between (i) *unresisted adiabatic expansion* and (ii) *adiabatic throttling*, and shew what functions of the state remain constant in these processes.

1 lb. of Ammonia (dry saturated) at 40° C. is caused to expand, according to the first method, down to a pressure of 125 lb. per sq. in. Shew that its dryness fraction is then 99·0 per cent.

Find the gain in entropy.

7. A steam engine working between limits of absolute temperature T_1 and T_2 follows a cycle similar to the Rankine with the exception of the adiabatic expansion, which is replaced by a stage in which the steam as it expands receives, per unit fall of temperature, heat at the rate of $k\frac{T}{100}-\left(\frac{T}{100}\right)^2$ per lb., k being less than $\frac{T_1}{100}$. If the steam at the beginning and end of expansion is just dry, determine k in terms of entropy and temperature and find an expression for the work done and the efficiency of the cycle. Find the value of the efficiency when $T_1=400°$, $T_2=300°$.

8. A triple-expansion engine is supplied with a feed heater which takes steam from the L.P. receiver. The admission temperature in the H.P. cylinder is 200° C., in the L.P. receiver is 95° C., in the condenser 45° C., and of the water before it enters the feed heater 25° C. Neglecting all effects of wire-drawing, clearance, drop, etc., and assuming the walls non-conductors of heat, compare the efficiencies

(1) when the feed heater is cut out,

(2) when the feed heater is supplied with 10 per cent. of the receiver steam.

PAPER XVIII

1. In an engine steam of dryness 0·95 is admitted to the cylinder at a pressure of 150 lb. per sq. in. and loses heat to the walls of the cylinder so that at the beginning of expansion it is of dryness 0·9. It receives heat during expansion so that the quantity received per unit fall of temperature is proportional to the absolute temperature, and it is just dry when it reaches

a pressure of 2 lb. per sq. in., at which pressure it is exhausted without further interchange of heat. Find the net heat received from the cylinder walls and determine the efficiency of the cycle.

2. In a compound engine with large receiver the pressures are:

In the High pressure cylinder at admission	120	lb. per sq. in.
,, ,, ,, ,, release	55	,, ,,
,, Receiver	30	,, ,,
,, Low pressure cylinder at release	6	,, ,,
,, Condenser	2	,, ,,

The steam is just dry at admission in the H.P. cylinder. Assuming that there is no loss of heat from the steam during expansion or in the receiver and neglecting the effect of clearance, calculate the work done per lb. of steam in each cylinder, and compare the total work done with that for the Rankine engine working between the same pressure limits. To what is the difference due?

3. In a steam-power plant, in which bleeding and reheating are both applied at the same interstage point at a pressure of 50 lb. per sq. in., the feed water is brought up from the temperature of the condenser to that of the higher stage exhaust, for which $I = 629$, and 50 TH.U. are available for reheating per lb. of the higher stage exhaust.

Assuming the plant to be operating on the Rankine cycle, compare the work done in the lower stage (1) by bleeding before, (2) by bleeding after, reheating.

[Condenser pressure: 0·8 lb. per sq. in.]

4. A compound steam engine is supplied with dry saturated steam at 200 lb. per sq. in. and the condenser pressure is 2 lb. per sq. in. The correct amount of steam to give best efficiency is bled at 25 lb. per sq. in. from the intermediate receiver. If the engine works on the Rankine cycle, determine the percentage reduction in volume of the L.P. cylinder resulting from bleeding for the same output.

5. Steam is supplied to an engine at 170 lb. per sq. in., 360° C., and expands to a dry saturated state at 20 lb. per sq. in. in the

H.P. cylinder. At exhaust from this cylinder the correct proportion to give best results is bled and the remainder reheated to 360° C. again, thereafter expanding to a dry saturated state at 1 lb. per sq. in. If the engine works on the Rankine cycle, determine its efficiency.

6. Sketch on the entropy-temperature diagram the following cycle proposed for a steam turbine. Give values of ϕ and T at the corners of the diagram.

Steam is admitted at a pressure of 200 lb. per sq. in. and temperature 400° C. ($I=780$). It expands isothermally in the turbine to a point from which an adiabatic expansion leaves it dry and saturated in the condenser at a pressure of 1·5 lb. per sq. in.

Determine the amount of heat added per lb. during the isothermal expansion. Shew that, as compared with simple superheating to 400° C. and an immediate adiabatic expansion to 1·5 lb. per sq. in., the efficiency is thereby increased by about 10 per cent.

If the isothermal and adiabatic expansion stages were to be carried out in separate turbines without loss of heat, shew that the work done in the turbines would be in the ratio 1 : 1·3 approximately.

7. In a compound engine in which the H.P. cylinder is of normal construction and the L.P. of uniflow type, steam is supplied at 200 lb. per sq. in. pressure and superheated to 300° C. Expansion in the H.P. cylinder is carried down to 40 lb. per sq. in. In passing to the L.P. cylinder throttling occurs, reducing the pressure to 30 lb. per sq. in. The condenser pressure is 2 lb. per sq. in. Assuming expansion in both cylinders is adiabatic and complete, find the loss of available energy per lb. of steam due to the throttling.

If 10 per cent. of the steam is extracted between the cylinders and mixed with the feed water instead of doing work in the L.P. cylinder, shew that the efficiency of the plant is improved.

What justification is there for the assumption of adiabatic expansion in such a combination?

8. A two-cylinder tandem compound double-acting engine, fitted with a large intermediate receiver, is supplied with steam at a pressure of 110 lb. per sq. in. and dryness 0·9. The condenser pressure is 3 lb. per sq. in. The total ratio of expansion is 6, the cut-off in the H.P. cylinder is at 0·4 stroke, the pressure at release in that cylinder is equal to the pressure (assumed constant) in the receiver, and the expansions follow the law $pv^n = \text{const}$. If the net maximum force due to steam pressure is the same on each piston, find the ratio of the cylinder volumes, the pressure in the receiver, the point of cut-off in the L.P. cylinder, the state of the steam at cut-off in the L.P. cylinder, and the value of n. Clearance, the areas of the piston-rods, and heat losses in the receiver are to be neglected.

If the boiler, condenser and receiver pressures, the state of the supply steam and the volumes of the cylinders remain unchanged, but the cut-off in the H.P. cylinder is adjusted so that the pressure at release in that cylinder is 50 lb. per sq. in., find the point of cut-off in the H.P. cylinder and the state of the steam at cut-off in the L.P. cylinder. The expansion in the H.P. cylinder follows the same law as in the first part of the question, and the cut-off in the L.P. cylinder is adjusted to satisfy the altered conditions.

PAPER XIX

1. The weight of air entering a condenser with the exhaust steam is 0·15 lb. per min. Find the volume of air and the weight of vapour that must be removed per min. from the condenser, the vacuum being 28 in., in the cases when the temperature in the condenser is (i) 60° C., (ii) 50° C. Barometer 30·1 in.

2. A boiler is delivering dry saturated steam at a pressure of 200 lb. per sq. in. abs. The feed water may contain as much as 4·4 lb. of air in 10^5 lb. of water. Shew that by neglecting the dissolved air we are only making an error of about 3×10^{-3} per cent. in the pressure of the steam.

3. An air pump for a condenser is to give a vacuum of 26 in. of mercury with a discharge temperature of 45° C., and it may be taken that the ratio of air to steam by weight in the exhaust pipe is 0·066. Assuming no leakage in the condenser and neg-

lecting clearance, slip, and leakage in the pump, find the volume swept through by the piston of the pump per unit volume of water discharged. Barometer 30 in.

4. In an engine air leakage into the system is 0·005 lb. of air per lb. of steam used. Determine the vacuum in inches of mercury in the condenser if the temperature there is 35° C. The steam entering the condenser has a dryness of 70 per cent.

Barometer 30 in.

5. Air completely saturated with steam at 60° C. is passed into a calorimeter at constant pressure and is cooled to 0° C. Calculate the quantity of heat given up per cu. ft. of the saturated air, assuming the constant pressure in the calorimeter to be 15 lb. per sq. in. abs.

6. In the flues of a boiler there are 0·5 lb. of steam to 20 lb. of dry flue gas, the pressure being 14·4 lb. per sq. in. Estimate the pressure of the steam.

It may be assumed that the steam is sufficiently highly superheated to be treated as a gas.

R for the dry gas is 94 in ft.-lb. units.

The molecular weight in lbs. of each of the gases occupies 358 cu. ft. at N.T.P.

7. A closed vessel contains 1 lb. of air and 3 lb. of water and steam at atmospheric pressure 15 lb. per sq. in. and temperature 15° C. Neglecting the effect of air actually dissolved in the water, find the heat required to raise the temperature to 150° C., and the resulting pressure.

8. The temperatures at entry to and exit from the steam side of a condenser are 50° C. and 35° C. respectively and the vacuum gauge reads 26·2 in. of mercury (barometer 30 in.). The heat carried away by the circulating water per lb. of exhaust steam is 393 C.TH.U. Shew that these figures are consistent with there being about 0·048 lb. of air present per lb. of steam.

The steam at exit may be neglected.

PAPER XX

1. A steam turbine condenser works with a vacuum of 28 in. of mercury (barometer 30 in. = 14·7 lb. per sq. in.). The temperature in the air ejector suction pipe is 36° C. If the ratio of air to steam, by weight, passing through the condenser is $3{\cdot}73 \times 10^{-4}$, shew that the air ejector has to remove about 0·64 cu. ft. of air in the condenser, per lb. of steam supplied to the turbine.

Shew that this volume would be diminished to about one-third the value, if the air was given a further cooling (with the same condenser vacuum) down to 30° C.

2. A boiler (at 10° C.) is half filled with water, half filled with air at 14·7 lb. per sq. in. (atmospheric pressure), the amount of water vapour in the air being negligible. With the stop and feed valves shut, the boiler is heated up till the contents are at 180° C. What pressure will be registered on the pressure gauge, assuming as a first approximation that there is no appreciable change in the volume of water present?

State approximately the weight of water that has been evaporated, and the amount of heat absorbed in the warming up of the boiler contents, estimated per lb. of water in the boiler. Neglect any solubility of the air constituents in water.

3. A vessel of volume 10 cu. ft. contains air at 180° C. and 40 lb. per sq. in. Find the weight of water which must be pumped in, so that when the vessel is heated, the temperature being maintained constant, the air will be just saturated with water vapour. How much heat must be supplied if the water temperature was initially 15° C.?

What quantity of heat must then be extracted from the contents of the vessel in order to reduce the temperature to 30° C.?

4. In a de-aerator for the water supply to a boiler, the feed water, warmed to 70° C., is sprayed into a chamber in which a partial vacuum is maintained: a part of the water is thereby evaporated and the vapour together with air is drawn off by an ejector. The main bulk of the water substantially freed from air is then pumped to the boiler.

If the pressure in the chamber is 23·5 in. mercury below atmosphere, and the temperature is 60° C., find the number of cu. in. of air at S.T.P. which could be withdrawn per lb. of feed entering the de-aerator.

[Barometer 30 in. = 14·7 lb. per sq. in.]

5. Dry hot air is passed into a cooling tower containing tiles which are kept wetted by a supply of water at 15° C. The air emerges saturated with water vapour at a temperature of 60° C. Shew that the quantity of water evaporated is about 0·15 lb. per lb. of air, and that the temperature of the entering air is about 440° C.

6. A steam turbine using 120,000 lb. of steam per hour exhausts into a condenser, in which the air-extractor suction-pipe pressure corresponds to a vacuum of 28·1 in. (barometer 30 in.) Hg, and the temperature in the same pipe is 33° C. The air leakage is estimated to be 0·9 lb. per 1000 lb. of steam passing through the turbine. The increase in temperature of the circulating water in passing through the condenser is 15° C. Estimate the volume in cu. ft. per min. passing to the air-extractor, and the quantity of circulating water required per min., if the dryness fraction of the exhaust steam from the turbine be 0·9.

7. In an evaporative surface condenser steam is condensed by the joint action of a film of cooling water and a supply of air and part of the cooling water evaporates. The unevaporated cooling water returns to the circulating pump where it mixes with make-up water and is conveyed back to the condenser inlet.

In a particular case the air supplied is fully saturated at 5° C. and at discharge is fully saturated at 20° C. Calculate the volume of air supplied per lb. of water evaporated, assuming the pressure throughout is 15 lb. per sq. in.

If the cooling water enters the condenser at 27° C. and returns to the pump at 38° C., and if the make-up water has a temperature of 5° C., estimate the air and water supply necessary to condense 10,000 lb. of steam per hr., supposing that 500 TH.U. are extracted per lb.

8. A wet-bulb thermometer, when properly wetted and placed in a stream of air of dry-bulb temperature 20° C., settles down

for a short time to a steady temperature of 15° C. Assuming that the air scrubbing the wet bulb is completely saturated (at 15° C.), shew that it then contains 74·2 grains of water vapour per lb. of dry air.

Write down a heat balance equation for the scrubbing process, considering m grains of water vapour per lb. of (dry) air in the air supplied to the wet bulb, and shew that $m = 60$ approximately. Neglect radiation, and assume standard barometric pressure.

[1 lb. = 7000 grains.]

PAPER XXI

1. The indicator diagram shewn is taken from a single-acting engine using 1200 lb. of steam per hr. and making 90 R.P.M. The clearance volume is 10 per cent. of the stroke volume. The stroke of the engine is 2 ft. and the diameter of the cylinder is 18 in. Determine the I.H.P. of the engine and find the dryness of the steam at the point marked A in the diagram. Atmospheric pressure = 15 lb. per sq. in.

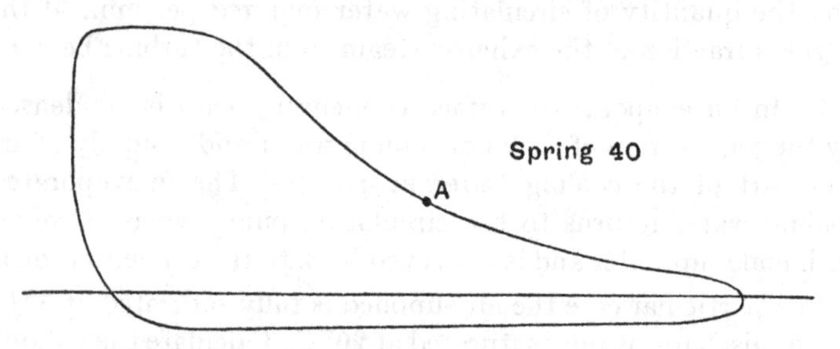

2. An indicator diagram, taken with a number 60 spring, is divided into ten strips of equal width and the lengths of the middle ordinates of these strips, measured in inches, are as follows: ·26, ·67, ·84, ·91, ·88, ·78, ·63, ·52, ·43, ·33. Find the mean pressure in the cylinder during one revolution.

The area of the piston is 27 sq. in., the length of the stroke is 5 in., and the engine is single-acting. Find the work done in one revolution, and the rate of working, in horse power, when the speed is 475 R.P.M.

3. The indicator diagram (180 spring) for a horizontal single-acting gas engine is given below. The diameter of the piston is 18 in., length of stroke 2 ft., speed 100 R.P.M. Find the I.H.P. if the engine is fully loaded, the speed of the engine being regulated by a cut-off governor.

If the clearance volume is $\frac{1}{3}$ the piston displacement, find the equations for the expansion and compression lines assuming they are of the form $pv^n = \text{const}$. Atmospheric pressure 15 lb. per sq. in.

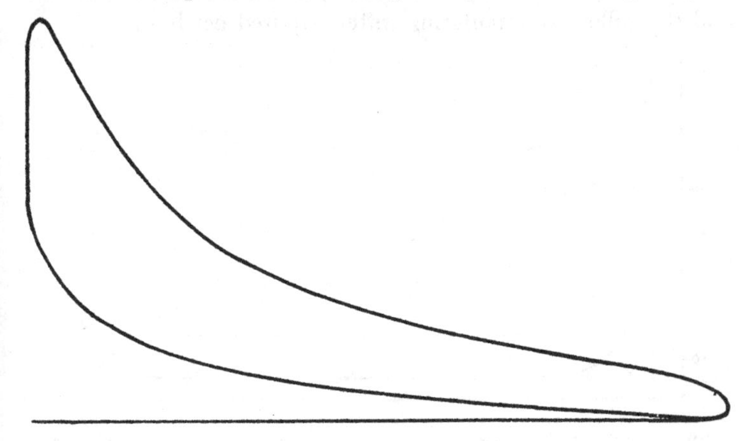

4. The indicator diagram is taken from the end of a single-acting cylinder of 1 ft. 6 in. diameter and stroke 2 ft. Determine the I.H.P. of the engine.

The clearance volume at either end of the cylinder is 10 per cent. of the stroke volume and the feed to the engine is 50 lb. per min. If the atmospheric pressure is 14·7 lb. per sq. in., determine the dryness of the steam at the point *A*.

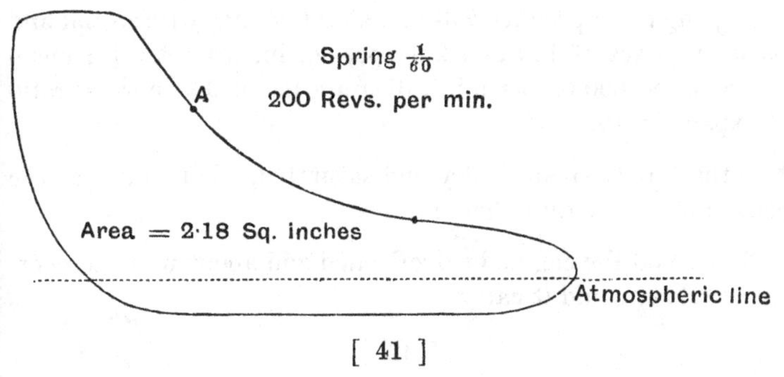

5. A double-acting engine has a stroke of 2 ft. and a cylinder diameter of 1 ft.; it runs at 96 R.P.M. Find its I.H.P. from the given indicator diagram.

The engine uses 17·3 lb. of steam per min.; if the steam supplied be taken as dry and if the temperature limits be taken as those corresponding to the maximum and minimum pressures, find the thermal efficiency. If the condenser be of the surface type and the temperature rise of the circulating water is limited to 40° C., find the gallons of circulating water required per hour.

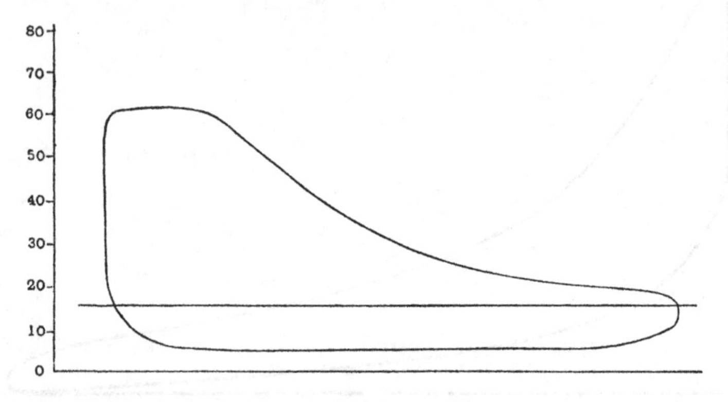

The engine has a clearance of 9 per cent. If the steam is dry at the end of compression, find the weight of cushion steam. Draw the saturation curve to scale on the diagram provided. Deduce the τ-ϕ line for expansion, and point out what it indicates as regards the heat exchanges between the steam and the cylinder walls.

6. Obtain the cylinder dimensions of a single-cylinder double-acting engine to produce 243 H.P. at 120 R.P.M., with initial and back pressures of 150 and 2 lb. per sq. in. respectively; mean piston speed 600 ft. per min.; diagram factor 0·8; normal ratio of expansion 10.

If the supply steam is dry and saturated, what would be the consumption per I.H.P. hour?

How could the engine be overloaded and about what percentage overload could it carry?

7. Determine the cylinder volume of a double-acting steam engine which may be expected to produce 250 I.H.P. at 125 R.P.M. Boiler and condenser pressures 200 and 2 lb. per sq. in. respectively. Ratio of expansion 12. Diagram factor 0·8.

Neglect clearance and assume hyperbolic expansion.

8. The stop-valve pressure of an engine is 100 lb. per sq. in. and the exhaust pressure 2 lb. per sq. in. With an expansion ratio of 5 and a mean piston speed of 750 ft. per min., determine the dimensions of the cylinder of a double-acting steam engine to develop 350 I.H.P. at 120 R.P.M. Use a diagram factor of 0·8.

PAPER XXII

1. The following observations were made in testing a jacketed engine, using steam of dryness 0·9 and pressure 200 lb. per sq. in. The jacket steam loses its latent heat to the cylinder but leaves at the temperature at which it enters.

Weight of steam used per minute 135 lb., of which 10 lb. goes to the jackets.

Indicated horse power, 500.
Condensing water per minute, 2050 lb.
Temperature of condensing water entering, 10° C.
Temperature of condensing water leaving, 40° C.
Temperature of air pump discharge, 50° C.

Draw up a heat balance sheet and find the amount of heat unaccounted for. Determine the thermal efficiency of the engine.

2. During a one-hour test of a condensing steam engine the following observations were made:

Pressure of steam at the stop valve	150 lb. per sq. in.
Dryness	0·95
Temperature of air-pump discharge	45° C.
,, condensing water entering	10° C.
,, ,, ,, leaving	35° C.
Total steam used	375 lb.
Condensing water used	7215 lb.
I.H.P.	25·4.

Draw up a heat balance sheet for the engine and determine the thermal efficiency.

3. A trial of a small compound double-acting non-condensing engine gives the following data:

Cylinder diameters, $5\frac{1}{4}$ in. and 9 in.
Stroke, 14 in.
Brake-wheel diam., 5 ft. Net load, 290 lb.
Mean effective pressure H.P., 53 lb. sq. in.
Mean effective pressure L.P., 17·5 lb. sq. in.
Revolutions per min., 140.
Feed-water, 450 lb. per hr.
Boiler pressure, 205 lb. per sq. in.
Exhaust pressure, 15 lb. per sq. in.

Determine the mechanical efficiency of the engine, the steam used per I.H.P. hour, and the amount of indicated work per lb. of steam.

Also find the amount of indicated work per lb. of steam in the case of a Rankine engine working between the same pressures.

4. Assuming no radiation and no leakage of steam or water in a steady-running engine, which is driving its own air pump, shew that the heat spent per min. in overcoming friction at the bearings and guides of the engine is 2900 TH.U. approx., when the weight of steam used per min. is 160 lb., the weight of condensing water per min. 3067 lb., the pressure of the boiler 115 lb. per sq. in. by gauge, the vacuum 27 in. of mercury, the temperatures of the condensing water at inlet and outlet 14° and 43° C. respectively, the B.H.P. of the engine 300 and the barometer reading 30·2 in.

5. A steam-jacketed engine working with steam at an initial temperature 120° C. indicates 50 H.P. and discharges 1250 lb. of water per hour to the hot well at a temperature of 50° C. The circulating water per hour is 20,000 lb. and its initial and final temperatures are 10° and 45° C. Neglecting radiation, find the amount of heat per min. received by the steam from the jackets; and assuming the jackets to be supplied with boiler steam, and only the latent heat of the steam to be given up to the working steam, find the jacket steam used per lb. of cylinder feed.

6. The following data are taken from the trial of a double-acting compound steam engine, running at a speed of 118 R.P.M.

Cylinder diameters H.P. $14\frac{3}{4}$ in., L.P. $25\frac{1}{2}$ in.
Stroke $29\frac{1}{2}$ in.
Mean effective pressure H.P. 63 and 66 lb. per sq. in.
L.P. 16 and 18 lb. per sq. in.

Calculate the I.H.P. of the engine, and its mechanical efficiency when delivering 295 H.P. at the flywheel.

In a trial of 9 hours the engine used 41,870 lb. of water, the pressure at the inlet valve being 142 lb. per sq. in. abs. and the temperature of the condensed steam 57° C. Find the thermal efficiency of the engine, and the weight of condensing water required per min. if its rise of temperature in the condenser is 9° C.

7. The following observations were made in a trial of a steam pumping plant:

Coal used per hour	456 lb.
Calorific value of the coal	8300 TH.U.
Feed water per hour	4522 lb.
Temp. of feed water	10° C.
Boiler pressure	75 lb. per sq. in.
Exhaust pressure	1 lb. per sq. in.
Water discharged from jet condenser per minute	2587 lb.
Inlet temp. of condenser water ...	10° C.
Outlet temp. of condenser water ...	26° C.
I.H.P.	255
Water pumped per minute	13400 gallons
Head of water in delivery pipe ...	54 ft.

Calculate the efficiency of the boiler, the thermal efficiency of the engine, the mechanical efficiency of combined engine and pump, and the efficiency of the plant as a whole.

Also draw up a balance sheet for the heat supply of the engine and estimate the amount of heat unaccounted for.

8. In a trial of a Worthington steam pump the following estimations were made from direct measurement:

Coal used per hour	=456 lb.
Estimated heat value of one pound of coal	=8000 TH.U.
Feed water per hour	=4522 lb.
Temperature of feed water	=15° C.
Boiler pressure	=75 lb. per sq. in.
Drainage from steam jacket per hour not returned to boiler, jacket supplied with steam of boiler pressure	=706 lb.
Condensing water, including condensed steam, discharged per minute	=2586 lb.
Initial temperature of condensing water	=15° C.
Final temperature of condensing water	=30° C.
I.H.P.	=255
Gallons of water pumped per minute	=13,400
Head of water in delivery pipe from pumps	=54 ft.

Calculate the efficiency of the boiler, of the steam in the engine, of the mechanism of the engine and pump and of the total pumping process. Also estimate the various quantities of heat which are not utilised and determine what amount of the total heat expended is not accounted for in the above observations.

PAPER XXIII

1. If the universal gas constant is 2776 in ft.-lb.-Centigrade units, determine the specific heats of marsh gas CH_4, the index for adiabatic expansion being 1·3.

Estimate the calorific value of CH_4 in C.TH.U. per standard cu. ft.

Heats of combustion:	Carbon	8,080 C.TH.U. per lb.
	Hydrogen	34,500 C.TH.U. per lb.
Heat of formation:	Marsh gas	1,275 C.TH.U. per lb.

2. A vessel contains a mixture of oxygen and hydrogen in the proportions 16 : 1 by weight at a pressure and temperature of 15 lb. per sq. in. and 15° C. respectively. If the mixture is exploded and all the hydrogen burnt, determine the pressure in the vessel when the temperature has fallen to 15° C. again.

3. Assuming that 1 cu. ft. of gas requires 7 cu. ft. of air for its combustion and that 1 lb. of the gas occupies 30 cu. ft. at pressure of 14·7 lb. per sq. in. and temperature of 15° C., find the pressure and temperature produced when 1 lb. of the gas at the given pressure and temperature is burnt at constant volume. The calorific value of the gas is 11,000 TH.U. per lb.

4. In a test of the calorific value of a coal by the Rosenhain calorimeter 1·7 grammes of coal are burnt in 2500 grammes of water and the water equivalent of the vessel is 500 grammes—the temperature of the air, water and vessel is initially 20° C. During combustion the temperature rises at a uniform rate to 23·5° C. in 5 min. and at the end of each of the two succeeding half-minutes is 23·75° C., falling afterwards at the rate of 0·25° C. per min. Find the thermal value of the coal in TH.U. per lb.

5. A sample of coal gives on analysis Carbon 81 per cent., Hydrogen 4·1 per cent., Oxygen 7·2 per cent. Find the amount of oxygen required for the complete combustion of one lb. of the coal, and the necessary air supply per lb. of coal.

Find also the gross and the net calorific values of the coal, taking the calorific value of carbon as 8060 TH.U. and that of hydrogen as 34,500 TH.U. per lb.

6. An analysis of coal gives Carbon 84 per cent., Hydrogen 5·6 per cent., Oxygen 7·4 per cent.: find the number of pounds of air required for the complete combustion of 1 lb. of the coal.

If the flue gases from the same boiler are found on analysis to contain, by volume, CO_2 12 per cent., CO 1·3 per cent., O_2 6·5 per cent., find the actual supply of air to the boiler per lb. of coal.

7. An analysis of coal gives Carbon 82 per cent., Hydrogen 5·4 per cent., Oxygen 7·1 per cent.: find the number of pounds of air required for the complete combustion of 1 lb. of coal.

Find also the theoretical calorific value of the coal if the calorific value of Carbon be taken as 8060 TH.U. per lb., and that of Hydrogen as 34,500 TH.U. per lb.

The flue gases from the furnace are found to contain by volume CO_2 12 per cent., CO 1·3 per cent. and O_2 6·5 per cent.: find the actual air supply to the furnace per lb. of coal.

8. The ratio of carbon to hydrogen in petrol corresponds to C_8H_{18}.

Determine the volumetric analysis of the dry exhaust gas, when

(*a*) petrol is burnt completely with 10 per cent. of excess air,

(*b*) when 10 per cent. of excess air is supplied, but the ratio of CO to CO_2 in the exhaust is 2 : 3 by volume and there is no unburnt hydrogen.

PAPER XXIV

1. The volumetric analysis of the dry flue gases from a boiler was CO_2 14 per cent., O_2 6 per cent., N_2 80 per cent. There were 16 lb. of dry flue gas and 0·6 lb. of steam per lb. of coal burnt. If the total pressure in the flue was 14·4 lb. per sq. in. and the temperature sufficiently high for the steam to be considered as a perfect gas, estimate the pressure of the steam.

2. A fuel gas consists of a mixture of CO and H_2 and when used in an engine the volumetric analysis of the dry exhaust gas is: CO_2 $8\frac{1}{3}$ per cent., O_2 $12\frac{2}{3}$ per cent., N_2 79 per cent. Determine the proportions of CO and H_2 in the gas and the volume of air supplied per cu. ft. of gas. Air may be taken to contain 21 per cent. of O_2 and 79 per cent. of N_2 by volume.

3. A mixture of coal gas and air (10 per cent. coal gas) is exploded in a closed vessel of 1 cu. ft. capacity. The initial temperature and pressure are 15° C. and 14·4 lb. per sq. in. respectively, and the maximum pressure attained in the explosion is 105 lb. per sq. in. The chemical contraction of the mixture is 2 per cent. and the calorific value of the coal gas is 300 TH.U. per standard cu. ft.

Estimate the average apparent specific heat at constant volume of the products of combustion in ft.-lb. per standard cu. ft. over the range of temperature of the explosion.

The latent heat of the steam may be neglected.

4. In the flues of a boiler, where the temperature is about 400° C., the dry flue gases consist of: CO_2 12 per cent., O_2 8 per cent., and N_2 80 per cent. by volume, and there is 0·05 lb. of steam per lb. of dry flue gas. If the pressure in the flues at this point is 14·6 lb. per sq. in., estimate the pressure of the steam.

Molecular weights: CO_2 44, O_2 32, N_2 28, steam 18. It may be assumed that the steam behaves as a perfect gas at the temperature stated.

5. A vessel, of volume 2·9 cu. ft., contains oxygen at a pressure and temperature of 15 lb. per sq. in. and 15° C. respectively. It is charged with hydrogen from a constant pressure supply at a temperature of 15° C., the resulting mixture being such that, on explosion, the two gases would just be completely united. Shew that the weight of hydrogen admitted is $\frac{1}{32}$ lb.

Assuming no heat interchange between the gases in the vessel and their surroundings, shew that the resulting temperature and pressure in the vessel is about 92° C. and 57 lb. per sq. in. respectively, γ for both gases being 1·4.

6. The products of combustion of gunpowder consist of gas with finely divided solid matter dispersed through it. The volume of the solid products is half that occupied by the powder, and their capacity for heat is 3 times the capacity for heat at constant volume of the gases formed from the same mass of powder. The value of γ for the gaseous products is 1·3. Shew that if the powder is fired in a chamber which it just fills, and the products of combustion are then expanded adiabatically, the pressure when the volume is 4 times the initial volume will be 0·123 of the initial pressure.

7. The volumetric analysis of a sample of coal gas is:

CO 20 per cent., CH_4 5 per cent., H_2 45 per cent., N_2 12 per cent., CO_2 18 per cent.

Determine the volumetric diminution per cubic foot of gas when the gas is burnt with a 50 per cent. excess of air and the resulting steam is condensed. Give also the volumetric analysis of the dry exhaust gas.

Air contains 20·9 per cent. by volume of oxygen.

8. A gas engine uses producer gas of the following composition (by volume):

CO	25
H	14
CO_2	6
N	55
	100·0

The exhaust gases contain 15 per cent. of CO_2. Shew that the ratio of the volumes of air to gas taken into the engine is 1·41. Air contains 21 per cent. of its volume of oxygen.

What percentage of oxygen would you expect to find in the exhaust gases?

PAPER XXV

1. Coal gas as supplied to a gas engine has a composition by volume as follows:

H_2 47 per cent., CH_4 35 per cent., CO 12·5 per cent., CO_2 0·5 per cent., N_2 5 per cent.

What ratio of air to gas by volume would just give complete combustion?

If the air : gas proportion is actually 9 : 1, calculate the percentage composition (by volume) of the cylinder contents during the expansion stroke and the ratio of the volumes of the charge measured at the same pressure and temperature before and after combustion.

2. The flue gases from a boiler are found to contain by weight CO_2 11·9 per cent., CO 0·15 per cent. and O_2 12 per cent. The boiler is being fired with coal which contains by weight 90 per cent. carbon, 6 per cent. hydrogen and 4 per cent. oxygen. Determine the calorific value of the fuel and the supply of air to the boiler per lb. of fuel. The calorific values of carbon and hydrogen are 8080 and 29,000 TH.U. respectively.

3. In a boiler trial the percentage analysis by volume of the dry flue gases from the boiler gave CO_2 9·3, CO 0·2, O_2 10·4, N_2 80·1. The percentage analysis by weight of the coal used was C 79·2, H_2 4·5, moisture 5, N_2 1·3, ash 10. Determine (i) the minimum weight of air required for the complete combustion of one pound of the coal, (ii) the weight of dry air being actually used in the boiler per lb. of coal burnt, (iii) the weight of steam passing away with the flue gases per lb. of coal.

The composition of air by weight is N_2 76·8; O_2 23·2; and the atomic weights are: C, 12; O, 16; N, 14; H, 1.

4. The induction system of an engine may be considered as consisting of a pipe through which air is flowing steadily and slowly and into which is injected the correct amount of liquid fuel for theoretically complete combustion. During the flow a quantity of heat is taken up from the walls of the pipe which would have been sufficient to have raised the temperature of the entering air to 130° C. at constant pressure. Neglecting the specific heat and vapour pressure of the fuel and assuming that the latent heat does not vary with temperature and that the whole of the fuel is evaporated before leaving the pipe, estimate the temperature of the mixture at exit from the pipe for each of the three fuels for which the particulars are given in the table.

If the pressure is standard throughout, determine also the heat which could be liberated by combustion per cu. ft. of mixture for each fuel.

	Calorific Value of Liquid per lb.	Air in lb. required for complete combustion per lb. of fuel	Latent heat
Petrol	10580	15·1	75
Benzol	9740	13·2	95
Alcohol	6590	9·0	225

5. In a small gas engine of 3 in. bore and $4\frac{1}{2}$ in. stroke measurements of the air and gas supply were made by the orifice method. From the following data calculate the mixture and ratio of volume supplied per stroke to stroke volume of cylinder, the revs. per min. being 300.

Gas supply. Diam. of orifice $\frac{3}{32}$ in. Head of water 1·8 in. Temp. of gas 8° C. Pressure of gas 1 in. of water. Density of gas ·038 lb. per standard cu. ft.

Contraction coefficient = 0·64. Barom. = 764 mm.

Air supply. Diam. of orifice $\frac{3}{8}$ in. Temp. of air 8° C. Head of water 1·5 in.

6. A mixture of carbon monoxide, with just sufficient air to burn it completely to CO_2, is exploded in a closed vessel of constant volume when initially at 15° C. The mean specific heats at constant volume of carbon dioxide and nitrogen up to a temperature t° C. may be taken as given by

CO_2 ...... $k_v = 7{\cdot}5 + {\cdot}0011\, t$ calories per gram molecule per ° C.

N_2 ...... $k_v = 4{\cdot}9 + {\cdot}00045\, t$,, ,, ,, ,, ,, .

while the heat of combustion of CO is 68 kilocalories per gram molecule.

Neglecting any loss of heat to the walls during combustion, calculate the maximum temperature reached on explosion, correct to 1 per cent. according to the data given.

What do you understand by the 'dissociation' which may occur at a high temperature? Shew how it will affect the maximum temperature reached.

Figures in calories per gram molecule may be converted to C.TH.U. per standard cu. ft., if desired, by multiplying by ·00283.

Ratio by volume of Nitrogen to Oxygen in air = 3·76.

7. Shew that the efficiency of a perfectly reversible heat engine working on the Carnot cycle does not depend on the nature of the working substance, but only on the temperature limits.

1 lb. of a fuel of calorific value H (in C.H. units) is burnt in a bomb (or constant volume) calorimeter with $(M-1)$ lb. of air, which are more than sufficient for complete combustion. Before burning takes place, both fuel and air are at atmospheric temperature T_0° C. abs. Assuming that the specific heat at constant volume of the products of combustion is $K_v = \alpha + \beta T$ at tem-

perature $T°$ C. abs., shew that the maximum temperature $T_1°$ C. abs. which can possibly be obtained is given by

$$\alpha (T_1 - T_0) + \frac{\beta}{2} (T_1^2 - T_0^2) = H/M,$$

and that the maximum mechanical energy that can be derived from the heat of combustion, with a lowest available temperature $T_0°$ C. abs., is

$$H - MT_0 \left[\alpha \log \frac{T_1}{T_0} + \beta (T_1 - T_0) \right].$$

8. The coal used in a boiler trial had the following composition: C 89·5 per cent., H 0·75 per cent., Ash 9·75 per cent. The dry flue gases contained, by volume, CO_2 10·10 per cent., CO 0·05 per cent., O_2 9·91 per cent., Nitrogen 79·94 per cent. Find the weight of air supplied per lb. of coal, the mean specific heat of the flue gases, and the heat carried away by the products of combustion and excess air, if the air temperature in the boiler house be 15° C., and that of the flue gases 320° C.

Specific heats: CO_2 0·216, CO 0·245, O_2 0·218, N_2 0·244, H_2O 0·480.

Atomic weights: H 1, C 12, O 16, N 14.

PAPER XXVI

1. Find the theoretical efficiency of an engine using the Otto cycle and drawing in air at a pressure of 15 lb. per sq. in. (1) when the air is compressed to 60 lb. per sq. in., (2) when it is compressed to 100 lb. per sq. in.

2. A gas engine has a stroke of 16 in. and its cylinder diameter is 9 in. The volume of the compression space is 240 cu. in. Find the ratio of compression, and the efficiency of an Otto cycle having the same ratio, when $\gamma = 1·4$.

The engine develops 25 I.H.P. on a consumption of 12·5 cu. ft., per min., of a gas with a calorific value of 315 thermal units per cu. ft. Find its thermal efficiency and the ratio of this to the corresponding Otto efficiency.

3. The indicated and brake horse powers of a gas engine are 15 and 12 respectively, and 4 standard cu. ft. of gas are used per minute, of calorific value 270 TH.U. The cylinder jacket takes

17·5 lb. of water per minute the temperature of which is raised from 10° to 28° C. An exhaust gas calorimeter takes 8 lb. of water per min. with a temperature rise of 40° C. The exhaust gases carry away from the calorimeter 10 per cent. of the heat supplied to the engine.

Draw up a heat balance sheet and determine the indicated and brake thermal efficiencies of the engine.

4. A Diesel engine is supplied with oil whose calorific value is 8000 TH.U. per lb. On test it was found that the I.H.P. = 44·6, B.H.P. = 30·2, oil per hour = 14·1 lb., circulating water 870 lb. per hour with a temperature rise of 40·5° C. Find the thermal efficiency, the mechanical efficiency, and estimate the heat lost in exhaust, radiation and friction.

5. In the trial of a gas engine the following observations were made:

Duration of trial 60 min., I.H.P. 55·9, B.H.P. 48·1, total gas used at standard temperature and pressure 880 cu. ft., lower calorific value of the gas in TH.U. per cu. ft. 310, jacket water per hour 980 lb., temperature of jacket water at entry 7·2° C., temperature of jacket water at exit 70° C., cooling water to exhaust calorimeter 4300 lb., temperature of water at entry to exhaust calorimeter 6·6° C., temperature at outlet 33·3° C., temperature of gas in the chimney 50° C. Make out a complete balance sheet shewing the distribution of the heat supply to the engine.

6. A gas engine in which the compression space is 0·2 cu. ft. and the stroke volume is 1 cu. ft. is using a mixture for which $\gamma = 1{\cdot}38$ and $k_v = 0{\cdot}18$. The temperature at the beginning of compression is 60° C. and the pressure 14·7 lb. per sq. in. What mass of the mixture is present in the cylinder?

The engine is working on the Otto cycle and the mixture absorbs 30 units of heat during the explosion; find the theoretical values of the pressure and temperature at the beginning and at the end of the expansion line.

7. Estimate the horse power developed by a motor car which travels 20 miles per gallon of petrol, at an average speed of

40 miles per hour. The specific gravity of the petrol is 0·8 and its calorific value 11,000 TH.U. per lb. The ratio of compression of the engine is 6 and the efficiency ratio (i.e. actual efficiency to air standard efficiency) is 0·65.

(1 gallon of water weighs 10 lb.)

8. The following data were obtained in a test of a Diesel engine:

Duration of test 60 min.

I.H.P. 12·95. B.H.P. 9·94.

Oil used 5·44 lb., calorific value (lower) 9340 lb.-cal. per lb.

Cooling water used 700 lb., rise of temperature 16·2° C.

Calculate the thermal efficiency relative to I.H.P., and to B.H.P., and find what percentage of the heat supplied is not accounted for in the above observations. What becomes of the heat which is not accounted for?

The air compressor is driven direct from the engine shaft and takes 1·1 H.P., of which 75 per cent. may be taken as usefully returned to the engine cylinder in the air supply. If this is taken into account, how does it affect the values of the thermal efficiency obtained above?

PAPER XXVII

1. In a gas engine working on the Otto cycle, the spontaneous ignition temperature of the gas is 470° C. and the temperature of the cylinder contents at the end of the suction stroke is 65° C. Determine the maximum ratio of compression which may be used and the theoretical maximum efficiency of the engine.

2. A six-cylinder four-stroke cycle Diesel engine of $13\frac{3}{4}$ in. bore and $14\frac{3}{4}$ in. stroke, with oil-cooled pistons, tested at half its normal output, gave the following results:

Room temperature: 18·6° C.

Barometer: 30·6 in.

R.P.M.: 350.

B.H.P.: 195.

Mean effective pressure (average): 53 lb. per sq. in.

Fuel: 101·5 lb. per hour.

Calorific value of fuel: 10,700 TH.U.

Jacket water: 171 lb. per min. with 31° C. rise in temperature.

Piston-cooling oil (sp. ht. = 0·5): 80 lb. per min. with 20° C. rise in temperature.

Exhaust calorimeter water: 148 lb. per min. with 27° C. rise in temperature.

Chimney gases: 35° C.

In the exhaust calorimeter, water was sprayed into the exhaust gases and it was estimated that 1140 cu. ft. of air per min. at engine room temperature and pressure (weighing about 88 lb.) entered the cylinders, molecular contraction in passing through the engine being negligible.

Draw up a heat balance sheet (in TH.U. per min.), indicating the items which may contain the various friction losses.

Also estimate the fuel consumption at full load in lb. per B.H.P. hour, if the efficiency of combustion is unchanged.

3. Sketch on a T-ϕ diagram for air, for the same range of temperature, lines representing

(i) heating at constant pressure,

(ii) compression according to the law $pv^n = \text{const.}$, where n is between 1·0 and γ,

and shew that the areas below these two lines represent

$$mK_p (T_1 - T_0) \quad \text{and} \quad \frac{mK_v (\gamma - n)}{n-1} (T_1 - T_0)$$

respectively, where T_0 and T_1 are the temperature limits. Hence shew that the sum of these areas represents the work done in a single stage air compressor.

Deduce, from this type of diagram, that for a two-stage machine, with the usual inter-cooling, the work done on the air should be the same in the two cylinders to give the greatest economy.

4. A four-cylinder four-stroke petrol engine is tested at constant speed and throttle opening. From the data below, estimate

the I.H.P. and indicated thermal efficiency, and compare the latter with the air standard efficiency.

B.H.P. with 4 cylinders working		18·8
,, ,, cylinder No. 1 cut out		12·9
,, ,, ,, 2 ,,		12·8
,, ,, ,, 3 ,,		12·6
,, ,, ,, 4 ,,		12·9
Fuel consumption	0·155 lb. per min.	
Calorific value of fuel	10,300 TH.U. per lb.	
Cylinder bore	69 mm.	
,, stroke	120 mm.	
Clearance volume	114 c.c.	

Briefly describe and justify any more accurate method of determining the I.H.P. of a four-stroke high speed petrol engine.

5. The clearance volume of a gas engine is one-third of the stroke volume, and at the end of the exhaust stroke is filled with exhaust gases at atmospheric pressure and at a temperature of 700° C. The charge of gas and air sucked in is also at atmospheric pressure and its temperature just before it enters the cylinder is 100° C. Assuming no loss of heat and that the specific heat of the exhaust gases is the same as that of the entering charge, find the temperature of the cylinder contents at the end of suction.

6. A petrol engine working on the Otto cycle has a cylinder 4 in. diameter by 5 in. stroke. The compression space is one-third of the stroke volume. At the end of the suction stroke the whole cylinder is filled with explosive mixture, at a pressure of 14·7 lb. per sq. in. at a temperature of 18° C. If 1 lb. of petrol requires 187 cu. ft. of air (at 0° C. and 14·7 lb. per sq. in.) for complete combustion and has a calorific value of 10,550 TH.U., find the maximum power which can be developed in the cylinder when running at 1000 R.P.M., assuming no losses.

7. An engine works on the Otto cycle and has a clearance volume of 0·3 cu. ft., and a piston displacement of 1·5 cu. ft. per stroke. At the beginning of compression the pressure in the cylinder is 14·7 lb. per sq. in. and the temperature is 50° C.

For the cylinder contents k_v may be taken as 0·18 and γ as 1·38. The engine takes 0·09 cu. ft. of gas per cycle, the calorific value of the gas being 320 thermal units per cu. ft. Find the mass of the cylinder contents, the pressure and the temperature at the end of compression and at the beginning of expansion (neglecting losses of heat), and the efficiency of the cycle.

8. In an air motor the upper pressure is 4½ atmospheres and the lower pressure 1 atmosphere. The clearance volume is 4 per cent. of the volume swept out by the piston and there is no compression. There is no loss of pressure between the air reservoir and the cylinder, and the temperature of the air in the reservoir is 64·4° C. The weight of air admitted per stroke is 0·07653 lb. and the expansion is adiabatic and complete. Estimate the weight of air shut in the clearance, the temperature at the point of cut off, the temperature at the end of expansion, and the work done per stroke.

PAPER XXVIII

1. The following data were obtained in a petrol engine test during 1 hour:

Mean I.H.P.	6·3
Effective circumference of brake drum and rope	11·25 ft.
Load on the brake ropes	48 lb.
Spring balance reading at other end of ropes	4 lb.
Mean speed	350 R.P.M.
Petrol used	2·89 lb.
Calorific value of petrol	10200 C.T.U. per lb.
Jacket cooling water	252 lb.
Jacket water temperatures	inlet 15° C. outlet 55° C.
Exhaust calorimeter water	210 lb.
,, ,, ,, temperatures	inlet 15° C. outlet 64·5° C.

Calculate: (1) The brake H.P.,
(2) Average torque in lb.-ft.,
(3) Mechanical efficiency,
(4) Brake thermal efficiency,
(5) H.P. lost in engine friction.

Draw up a complete heat balance and determine the percentage of heat supply unaccounted for.

Discuss briefly the question whether the above calorific value should be the 'higher' or the 'lower'.

2. A single cylinder, single acting gas engine working on the Otto cycle is to develop 56 I.H.P. at 240 R.P.M. From the following particulars estimate the swept volume of the cylinder.

Compression ratio = 5.

Air to gas ratio = 5.

Volumetric efficiency = 80 per cent.

Ratio of indicated thermal efficiency to air standard efficiency = 0·60.

Calorific value of gas = 280 TH.U. per standard cu. ft.

Engine room temperature = 15° C.

Engine room pressure = 14·7 lb. per sq. in.

γ for air = 1·4.

3. The volumetric efficiency of a gas engine working on the 'four-stroke' cycle is 0·82, the compression ratio being 5 and the prevailing engine room conditions 15° C. and 15 lb. per sq. in. If the residual gases are at a temperature of 600° C. and a pressure of 15 lb. per sq. in. and the pressure in the cylinder at the end of the suction stroke is 12 lb. per sq. in., estimate the temperature of the charge at that point.

If the law of compression is $pV^{1.30}$ = const., estimate also the temperature and pressure in the cylinder at the end of the compression stroke.

It may be assumed that only air is being considered.

4. A gas engine, working on the four-stroke cycle and running at 240 R.P.M. has a cylinder 9 in. bore and 18 in. stroke. It is

supplied with 7 standard cu. ft. of gas and 56 standard cu. ft. of air per min. when fully loaded. The compression space is 191 cu. in., the room temperature 17° C. and the calorific value of the gas is 280 TH.U. per standard cu. ft. If the thermal efficiency may be taken as 0·65 of that of the air standard cycle, estimate the volumetric efficiency of the cylinder, and the probable I.H.P. of the engine. Atmospheric pressure 15 lb. per sq. in.

5. The cylinder of a petrol engine is 4 in. in diameter and the stroke 5 in. The compression space is one-quarter of the stroke volume. At the end of the suction stroke the whole cylinder is filled with a mixture of air and petrol at a pressure of 14·7 lb. per sq. in. and temperature 80° C. If 1 lb. of petrol requires 180 standard cu. ft. of air for complete combustion and has a calorific value of 10,000 TH.U., find the maximum power which can be developed in the cylinder when running at 1000 R.P.M., assuming no losses.

6. 1 cu. ft. of air at atmospheric pressure and 278° C. is mixed with 4 cu. ft. of air at the same pressure and 27° C., the total volume remaining unaltered. Neglecting losses, shew that the resulting temperature of the mixture is 82° C., and that the pressure of the mixture increases by about 1 per cent. of an atmosphere. Assume at T° C. (abs.)

$$K_v = (4{\cdot}50 + 0{\cdot}0010T)/28{\cdot}85.$$

7. In a four-stroke gas engine the stroke volume is 1·25 cu. ft., the clearance volume 0·235 cu. ft., and release occurs at 0·89 of the stroke. When fully loaded, the engine takes in 1·06 cu. ft. of mixture at 15° C. and at atmospheric pressure 14·7 lb. per sq. in. If the pressure at release is 50 lb. per sq. in., find the temperature of the cylinder contents at the beginning of the compression stroke. Assume that the release valve is closed before the admission valve opens, and that $\gamma = 1{\cdot}35$.

8. A gas engine has a cylinder diameter 11·5 in., stroke 21 in. and a compression ratio 6·37. The volume of mixture taken in per cycle is found to be 0·83 times the stroke volume. The outside temperature is 15° C. and the pressure 14·7 lb. per sq. in. At the beginning of suction the temperature of the burnt products is 900° C., and for these $\gamma = 1{\cdot}25$ and the specific heat at constant pressure 0·31. For the mixture entering, the specific

heat at constant pressure is 0·242 and $\gamma = 1{\cdot}4$. Assuming the pressure during suction is atmospheric, find the heat supplied to the gases from the inlet valve and cylinder walls.

PAPER XXIX

1. A single-acting reciprocating air engine, with a stroke of 6 in. and piston diameter 4 in., is being driven from a reservoir of compressed air of 10 cu. ft. capacity; the pressure of the reservoir falling as the air is used in the engine. At the beginning of a particular cycle of the engine, the reservoir pressure is 120 lb. per sq. in. and the temperature 15° C. Find the work done in this cycle if the engine cuts off at $\frac{1}{4}$ stroke and exhausts into the atmosphere. Neglect clearance and assume the process adiabatic.

Find also what will be the temperature of the reservoir after the cycle.

Pressure of atmosphere 14·7 lb. per sq. in.

2. A gas engine developed $38\frac{1}{2}$ B.H.P. and indicated 45. It used 55 cubic feet a minute of gas at 21° C. and with a barometric pressure of 30·7 in.: the water used in the jacket was 65 lb. per min., the rise of temperature being 22·2° C. The calorific value of the gas at 15·5° C. and 30 in. barometer was 78 TH.U. per cu. ft. Draw up a heat balance sheet as far as the data permit, and find the mechanical and thermal efficiencies.

How would you obtain a rough estimate of the temperature at release?

3. In the trial of a gas engine the following observations were made:

Duration of trial 60 min., I.H.P. 60, B.H.P. 52, total gas used 950 cu. ft., total air used 8000 cu. ft., both being at standard temperature and pressure; higher calorific value of the gas 360 TH.U. per standard cu. ft., jacket water per hour 1060 lb., temperature of jacket water at entry 12° C., at exit 75° C., cooling water to exhaust calorimeter 5000 lb., temperature of water at entry to exhaust calorimeter 12° C., at exit 36° C., temperature of gas in the exhaust pipe 50° C. The specific heat of the gases is 0·24 and the temperature of the gas and air supply is 15° C.

Make out a complete balance sheet shewing the distribution of the heat supply to the engine.

4. A gas engine takes in 4·3 standard cu. ft. of gas and 35·5 standard cu. ft. of dry air per min. The exhaust gases from the engine are cooled in a calorimeter and then pass into the exhaust pipe at a temperature of 40° C. and a pressure of 14·85 lb. per sq. in. The temperature of the atmosphere is 15° C. Assuming that the gases are fully saturated, estimate the heat carried up the exhaust pipe. State what approximations you make in this estimate.

5. In a gas engine the temperature of the exhaust gases left in the clearance space at the end of the exhaust stroke is 650° C. and the temperature of the suction charge of gas and air just before it enters the cylinder is 100° C. The clearance volume is one-third of the piston displacement volume. Find the temperature of the gases in the cylinder at the end of the suction stroke, assuming that the exhaust gases and the suction charge have the same constant specific heat.

Why is it that the indicator diagrams taken from a gas engine running light are larger than those taken from the same engine when running under a load?

6. In the trial of a gas engine the following observations were made:

Duration of trial 60 min., I.H.P. 55·9, B.H.P. 48·1, total gas used at standard temperature and pressure 880 cu. ft., lower calorific value of the gas in TH.U. per cu. ft. 310, jacket water per hour 980 lb., temperature of jacket water at entry 7·2° C., temperature of jacket water at exit 70° C., cooling water to exhaust calorimeter 4300 lb., temperature of water at entry to exhaust calorimeter 6·6° C., temperature at outlet 33·3° C., temperature of gas in the chimney 50° C. Make out a complete balance sheet shewing the distribution of the heat supply to the engine.

7. During the trial of a gas engine and producer plant the following observations were made:

Analysis of the exhaust gas from the engine			
CO_2	...	15·5	per cent.
O_2	...	4·9	,, ,,
CO	...	0·0	,, ,,
N_2	...	79·6	,, ,,

Simultaneously with the exhaust gas sample, a sample of the gas from the producer was taken, 50 c.c. of which were mixed with 52 c.c. of air and exploded.

The volume of the mixture after explosion was ...	87·3 c.c.
,, ,, ,, absorbing the CO_2 was	71·5 ,,
,, ,, ,, ,, O_2 ,,	69·5 ,,
,, ,, ,, ,, CO ,,	69·5 ,,

Calculate the ratio of air to gas used by the engine.

8. An engine works on a theoretical Diesel cycle (with constant pressure combustion and constant volume rejection of heat) except that owing to heat exchanges between the cylinder walls and working fluid the adiabatics are replaced by curves of the type $pV^n = \text{constant}$, where $n < \gamma$ and is the same for both compression and expansion. The increase of weight of the charge due to fuel injection is to be neglected. Shew that if the compression ratio be r, the cut-off ratio R, and p be the compression pressure (abs.), the mean effective pressure is given by

$$\frac{pn(R-1)}{(n-1)(r-1)}\left[1 - \frac{R^n - 1}{n(R-1)\,r^{n-1}}\right].$$

If the compression be from 14·7 to 470 lb. per sq. in. (both pressures absolute), $n = 1{\cdot}35$, $R = 2{\cdot}4$, shew that the mean effective pressure is about 109 lb. per sq. in.

What are the chief factors limiting the mean effective pressure attainable in a Diesel engine?

PAPER XXX

1. An engine of 12 in. bore, 18 in. stroke, working on the Otto cycle, has a compression ratio of 7, the mass of the cylinder contents being 0·085 lb. During expansion the pressure falls from 600 lb. per sq. in. to 50 lb. per sq. in., the mean pressure during this period being found to be 125 lb. per sq. in. The properties of the working fluid are: Gas constant $R = 99{\cdot}4$ (lb.-ft. units) and $K_v = 0{\cdot}17 + 4{\cdot}5 \times 10^{-5}\,T$ TH.U. per lb. (T being absolute temperature).

Find the loss of heat to the walls during one expansion.

Assuming a value of K_v equal to the mean of the values at the beginning and end of expansion, find what the mean pressure would have been if there had been no loss of heat.

2. An aero-engine, fitted with a supercharger, is working at an altitude where the atmospheric pressure is 9 lb. per sq. in. The exhaust valve closes at the end of the stroke, leaving the compression space filled with exhaust gases at 800° C.: the suction valve then opens and the supercharger supplies fresh charge at 10° C. and at a uniform pressure of 13 lb. per sq. in. during the charging stroke. If the compression ratio is 5 and the stroke volume $\frac{1}{10}$ cu. ft., find the temperature of the cylinder contents at the end of the stroke and the mass of fresh charge entering. Assume that both the incoming charge and the exhaust gases have the same specific heat as air and that it is independent of temperature. Neglect any valve losses and heat received from the walls.

3. An aero-engine of compression ratio 6 to 1 is under test in an 'altitude chamber' where the pressure is 10 lb. per sq. in. and the temperature 15° C. The engine exhausts into a receiver where the pressure is the same as that in the chamber.

The volume of fresh charge drawn in during each suction stroke when reckoned at the pressure and temperature of the chamber is 90 per cent. of the swept volume.

Assuming that the mean specific heats at constant pressure of the residual gas and of the fresh charge are 0·3 and 0·24 respectively, and that the densities of the two gases at N.T.P. are the same, calculate the temperature of the cylinder contents at the end of the suction stroke, taking the temperature of the residual exhaust gas as 800° C.

What error is produced in the calculated temperature by an error of 100° in the estimated exhaust gas temperature?

4. Shew that the efficiency of an internal combustion engine made to given drawings depends on the size and on the speed in a manner given approximately by the expression $1-\frac{A}{D}-\frac{B}{nD}$, where D is the cylinder diameter, n the number of revolutions

per minute, and A and B are constants. State the chief factors in the types of loss represented by A and B respectively.

Shew that, if the piston speed be given, the efficiency is proportional to $1-\frac{\alpha}{D}$, where α is constant; and hence prove that the engine of least weight per horse power will have cylinders of diameter 2α.

5. In the combustion space of an internal combustion engine working on the ideal Otto cycle are placed a number of thin sheets of metal which may be assumed to follow the temperature of the working fluid. The mass of the metal is M_1 and the specific heat (assumed constant) K. The mass of the working fluid is M and the specific heats (assumed constant) at constant pressure and volume are K_p and K_v respectively. Shew that, assuming no heat losses and that the working fluid behaves as a perfect gas, the law of compression is $pV^n=\text{constant}$, where

$$n=(MK_p+M_1K)/(MK_v+M_1K).$$

Shew that if the circumstances (including compression ratio) are so arranged that the temperatures at the beginning and end of compression have the same values before and after fitting the sheets, then the theoretical efficiency of the engine will be the same in the two cases. Shew, however, that for equal charges of working fluid, the temperature rise during the explosion period will be less with than without the sheets fitted.

6. It is found that the area of the diagrams of the Crossley engine in the laboratory averages about 20 per cent. bigger when the engine is running light than when it is fully loaded. Explain this.

When the engine is running fully loaded the temperature of the exhaust gases left in the clearance space at the end of the exhaust stroke is 700° C., and the temperature of the gas and air sucked in just before they enter the cylinder is 100° C. The clearance space is $\frac{1}{4}$ of the total cylinder volume (including clearance space). Shew that the temperature of the gases filling the cylinder at the end of the suction stroke will be 170° C. Assume that no heat is lost to or gained from the cylinder walls during suction, that the pressure inside the cylinder is the same as that of the atmosphere, and that the specific heat of the exhaust gases, and of the incoming charge, is the same constant quantity.

7. An air-pump working in an air-ship takes air direct from the atmosphere where the pressure is 10 lb. per sq. in. The inlet valve closes at the completion of the suction stroke, and the pressure is then just equal to that of the atmosphere. The mean pressure in the pump during the suction stroke is 2 lb. per sq. in. below atmosphere. Neglecting clearance, and assuming that no heat passes between the cylinder walls and the cylinder contents during the suction stroke, shew that the volume of air (reckoned at external temperature and pressure) taken per stroke is 5·7 per cent. less than the stroke volume. The volumetric heat of air may be taken as 19·5 ft.-lb. per standard cu. ft.

8. The figure shews a light spring diagram of a four-stroke gas engine, the compression ratio of which was 5·5. If the volumetric efficiency, referred to S.T.P. conditions, was 78 per cent. and the temperature of the residual gas at the commencement of the suction stroke 600° C., find the mean temperature of the gases in the cylinder at that point of the compression stroke where the pressure was atmospheric. The specific heats of the charge and the residual gases are to be assumed to be equal.

If the temperature of the gas and air entering the suction pipe was 20° C. and the pressure 14·7 lb. per sq. in., estimate the amount of heat per lb. of incoming charge taken up from the walls during the period in which the pressure in the cylinder was below atmospheric.

(For mixture of gas and air $K_p = 0{\cdot}25$, specific volume at S.T.P. = 13 cu. ft. per lb.)

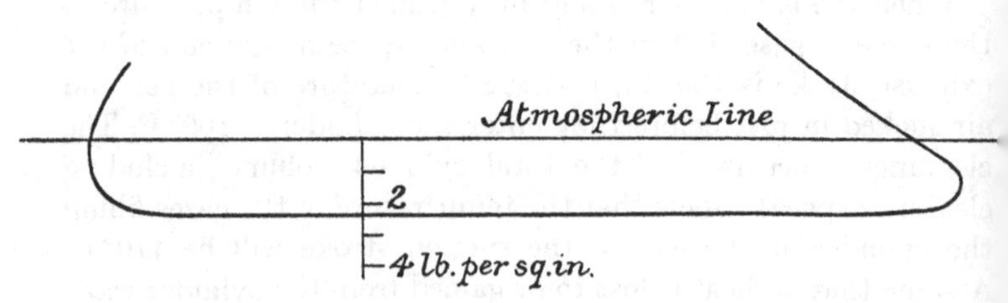

PAPER XXXI

1. A boiler works under a pressure of 200 lb. per sq. in. and evaporates 10 lb. of water per lb. of coal. The feed water passes through an economiser placed in the path of the furnace gases, and these gases have their temperature reduced from 320° C. to 210° C. in consequence. The air supply to the furnace is 20 lb. per lb. of coal, the atmospheric temperature is 15° C. and the specific heat of the furnace gases is 0·24. The feed water enters the economiser at a temperature of 36° C. Calculate the rise of temperature of the feed water in the economiser and estimate the calorific value of the coal, assuming that 90 per cent. of it is accounted for above.

2. A boiler is supplied with 200 lb. of air per min. at 20° C., dew point being 10° C. If the temperature of the flue gases leaving the boiler is 300° C., estimate the heat carried away per min. by the moisture from the air. Atmospheric pressure 14·7 lb. per sq. in.

If the coal contains 10 per cent. by weight of moisture and 4 per cent. by weight of available hydrogen and if 12 lb. of coal are consumed per min., determine the total weight of steam passing through the flues per min., and estimate the heat carried away by it.

3. A boiler trial supplies the following data: Steam pressure 180 lb. per sq. in., dryness fraction of steam 0·98, temperature of feed water 50° C., feed water 9·3 lb. per lb. of coal, temperature of chimney gases 270° C., temperature of air supplied 15° C. An analysis of the coal gave 87 per cent. carbon, 4 per cent. hydrogen, 3 per cent. oxygen by weight. Determine the calorific value of the fuel, the efficiency of the boiler, and the percentage loss of heat to the chimney, assuming 25 lb. of products per lb. of coal, and specific heat of products equal to 0·24.

4. In the trial of a water-tube boiler lasting 6 hours the following observations were made:

Coal consumed, 42,524 lb.
Calorific value of coal, 6540 TH.U.

Steam pressure, 200 lb. per sq. in.
Steam temperature, 325° C.
Water evaporated, 329,296 lb.
Temperature of feed to boiler, 108° C.
Weight of flue gas per lb. of coal, 18 lb., of which 0·5 lb. is steam.
Specific heat of dry flue gas, 0·24.
Temperature of flue gases leaving the boiler, 350° C.
Temperature of outside air assumed dry, 15° C.

Determine the efficiency of the boiler and the heat carried away per min. in the flue gases.

5. In the trial of a boiler and economiser the following observations were made:

Feed water per lb. of coal	7·25 lb.
Steam pressure	150 lb. per sq. in.
,, temperature	250° C.
Temperature of feed to boiler	130° C.
,, ,, economiser ...	35° C.
,, flue gases leaving boiler	380° C.
,, ,, ,, economiser	185° C.
,, boiler house	15° C.
Ash from grate per lb. of coal	0·06 lb.
Calorific value of ash per lb.	1500 TH.U.
,, ,, coal per lb.	6800 TH.U.

19·35 lb. of dry flue gas and 0·51 lb. of steam are produced per lb. of coal burnt. There is no CO in the flue gas. The specific heat of dry flue gas is 0·24.

Draw up a heat balance sheet, estimated on 1 lb. of coal, for (i) the boiler, (ii) the economiser.

6. In a trial of a Lancashire boiler and economiser the following figures were obtained:

Atmospheric pressure	14·7 lb. per sq. in.
Boiler room temperature	18° C.
Boiler room dewpoint	10° C.
Temperature of gases leaving furnace ...	420° C.
Temperature of gases leaving economiser	200° C.

Air consumption, dry air per lb. of coal 18 lb.
Calorific value of coal ... ... 7500 TH.U. per lb.

The analysis of the coal gave:

Water	5	per cent.
Ash	10	,,
Available hydrogen	5	,,

Determine the percentage of the heat supply carried away in the flue gases (*a*) leaving the furnace, (*b*) leaving the economiser.

State the justification for the use of the steam tables in this connection.

[K_p for dry flue gas 0·25.]

7. The following data refer to the test of a steam boiler with air pre-heating:

Coal.

Weight per hour = 3000 lb.
Calorific value = 6900 per lb. of *dry* coal.
Analysis by weight: carbon 70 per cent., available hydrogen 4 per cent., moisture 4 per cent.

Steam.

Weight per hour = 25,500 lb.
State of steam = 400 lb. per sq. in., 400° C. (I = 775·3).
Feed water = Temp. to boiler 173° C.

Air.

Volume per hour = 768,000 cu. ft. at 15 lb. per sq. in., 15° C.
Dewpoint low enough for moisture in air to be neglected.
Pre-heat to 215° C. before entering furnace.

Chimney gases.

Temperature at entry to pre-heater = 450° C.
Temperature leaving pre-heater = 250° C.
Specific heat at constant pressure of dry gases = 0·25.

Draw up a heat balance sheet on the basis of 1 lb. of coal (as used), and determine the over-all efficiency of the boiler and the percentage of the heat of combustion saved in the pre-heater.

8. In the published results of a boiler test, the fuel and flue gas analyses are given as follows:

	Coal analysis by weight %		Flue gas analysis by volume %
Carbon	83·44	CO_2	11·7
H	3·99	O	3·85
O	2·88	CO	0·45
Ash	9·69	N	84·0

Are these figures consistent with reasonably complete combustion of the coal, and how could they be explained? The atomic weight of carbon is 12, that of oxygen is 16, and the ratio of the nitrogen to the oxygen in the air is 3·76 by volume.

PAPER XXXII

1. A compressor takes in 2·5 cu. ft. of air at 14·7 lb. per sq. in. and at a temperature of 20° C. and compresses it isothermally to a volume of 0·25 cu. ft. The air is then cooled, at this volume, to a temperature of 12° C. and is then expanded adiabatically until its volume is 1·25 cu. ft. Find the pressure (1) at the end of compression, (2) at the end of cooling, (3) at the end of adiabatic expansion.

Also plot the p-V and T-ϕ diagrams for the whole process.

2. A double-acting air compressor makes 60 R.P.M.: it takes in air at 15 lb. per sq. in. and 150° C. and compresses it to 75 lb., when a valve opens and lets the air into a large chamber. The piston sweeps through 5 cu. ft. in its stroke, and the end clearances are each 5 per cent. of the active stroke. Sketch in roughly to scale the indicator diagram if the compression and expansion curves are of the form $pv^{1\cdot3}$ = constant. Also find the temperature of the air as admitted to the reservoir, the volume of air taken in per stroke, and the H.P. used in compression.

3. Air is drawn into a compressing cylinder at atmospheric pressure (14·7 lb. per sq. in.) and temperature (15° C.). It is compressed adiabatically to a pressure of 120 lb. per sq. in. and is delivered at this pressure to supply mains in which it is cooled to atmospheric temperature. The air is drawn from the mains

at the pressure of 120 lb. and expands adiabatically, in an air motor, down to atmospheric pressure. Find the work done, per lb. of air, in a complete cycle, in the compressing cylinder and the motor cylinder respectively, and the efficiency of the transmission.

4. A single-stage air compressor, in which the stroke volume is V and the clearance volume kV, is delivering air into a receiver of volume mV. If compressions and expansions follow the law pV^n = constant and atmospheric pressure is p_0 and suction pressure αp_0, shew that the apparent volumetric efficiency for that particular cycle, in which delivery pressure commences at xp_0, will be

$$(1+k).\alpha^{\frac{1}{n}} - kx^{\frac{1}{n}}.$$

Find the pressure in this cycle. Cooling in the receiver during a cycle may be neglected.

5. Adiabatic compression of air from pressure p_1 to pressure p_2 in one stage is replaced (i) by compression in two stages, (ii) by compression in three stages. Find in each case the values of the intermediate pressures which give the greatest economy of work in the compressor.

If the limiting pressures be 15 lb. per sq. in. and 150 lb. per sq. in., find the amount of work required per lb. of air compressed in each of the three cases.

Find also the amount of heat extracted from the air in each case by the constant pressure cooling.

6. The low-pressure cylinder of a two-stage air compressor works between the limits of pressure p_1 and p_2, and the high-pressure cylinder between p_2 and p_3. If the air is cooled at the intermediate pressure to its initial temperature, and if the compression in both cylinders is in accordance with the law pV^n = constant, shew that the work done per lb. of air will be a minimum when $p_2^2 = p_1 p_3$. Clearance is to be neglected.

Also shew that, with this condition satisfied, the whole operation is equivalent to a compression according to the law $pV^{\frac{2n}{n+1}}$ = constant, and that the ratio of the work required without intercooling to that with intercooling is $\frac{1}{2}(r^{\frac{n-1}{2n}}+1)$, where r is the ratio of the final to initial pressure.

7. What is the ratio of the efficiencies of two air compressors, compressing air from 15 to 150 lb. per sq. in., one in a single stage, the other in two stages, of equal pressure ratios? The compression follows the law $pV^{1\cdot 3}=$ a constant. The air is cooled to its initial temperature between the stages in the second case. Neglect clearance and mechanical losses.

8. An air compressor takes in air at atmospheric pressure, 14·7 lb. per sq. in., and at a temperature of 20° C., and compresses it to a pressure of 80 lb. per sq. in. on the gauge. If 200 lb. of air are dealt with per min., what power is absorbed? The law of compression is $pV^{1\cdot 3}=$ a constant.

What proportion of the power is lost if the air on leaving the compressor is allowed to cool, at constant pressure, to its initial temperature?

PAPER XXXIII

1. A small reservoir for compressed air has a volume of 2 cu. ft., and is being charged from the atmosphere by means of a single-acting pump of diameter 6 in. and stroke 9 in. At the beginning of one cycle of the pump, the reservoir pressure is 20 lb. per sq. in. by gauge. Find the pressure after one cycle of the pump, neglecting clearance and assuming that the air remains at the temperature of the atmosphere, 15° C. The atmospheric pressure is 14·7 lb. per sq. in.

Find also the work done by the pump and the mass of air pushed into the reservoir.

2. Air at a pressure of 15 lb. per sq. in. and temperature 20° C. is compressed adiabatically to a pressure of 150 lb. per sq. in. and delivered at this pressure to supply mains in which it is cooled to its initial temperature. The air is drawn from the mains and expanded adiabatically in an air-motor down to atmospheric pressure. Determine, neglecting clearance, (i) the work done in the compressor per 1000 standard cu. ft. of air, (ii) the efficiency of the transmission.

3. A single-acting air compressor supplying air to mains at 10 atmospheres pressure is driven by an engine giving 10 B.H.P. The air enters the compressor at 15° C. and atmospheric pressure, and is compressed adiabatically in one stage. If the mechanical efficiency of the compressor is 85 per cent. and the speed is 200 R.P.M., how many lbs. of air can the plant supply per hour, and what must be the size of the pump cylinder? Losses due to clearance etc. may be neglected.

By how much would the lbs. per hour be increased if the compression could be made isothermal instead of adiabatic?

4. Distinguish between the terms *true* and *apparent* volumetric efficiency as applied to an air compressor. Indicate briefly how you would measure these items in practice.

An air compressor has a clearance space of α times the stroke volume. The indicator diagram is such that the delivery and suction lines are lines of constant pressure, the pressures being p_2 and p_1 respectively. The value of p_1 is however less than that of the atmospheric pressure p_0. Assuming that compression and expansion follow the same law $pV^n = \text{constant}$, shew that the apparent volumetric efficiency is

$$(1+\alpha)(p_1/p_0)^{\frac{1}{n}} - \alpha (p_2/p_0)^{\frac{1}{n}}.$$

Evaluate this expression when $p_2=105$, $p_1=14$, $p_0=15$ (all pressures in lb. per sq. in.), $\alpha=0{\cdot}02$, $n=1{\cdot}2$.

5. An air compressor (turbo-compressor) under test gave the following data:

Barometer 30 in.; Hg = 14·7 lb. per sq. in. = suction pressure.
Delivery pressure = 80·0 lb. per sq. in.
Temperature of suction air = 20·5° C.
" " delivery " = 79·5° C.
Jacket cooling water per hour = 8680 gallons.
Temp. of " " (entry) = 23·62° C.
" " " (exit) = 34·00° C.
Quantity of air per hour = 324,000 cu. ft. under prevailing atmospheric conditions.

Make out a heat balance sheet in C.H. units per min., and find the shaft horse power (H.P.) of the turbo-compressor. Find the theoretical H.P. required to compress isothermally the given quantity of air per hour, and state the ratio (the isothermal compression efficiency) of this to the actual shaft H.P. Neglect radiation and bearing friction.

The air may be treated as dry. State briefly (without formulae) how the air quantity per hour could be measured.

6. What is the object of arranging for compressing air in stages with intermediate water-cooling down to the original atmospheric temperature? Give the usual rules for the values of the intermediate pressures in the case of (i) two-stage, (ii) three-stage, compression, and explain the circumstances in which these rules hold.

A three-stage air compressor (with inter-coolers) for a Diesel engine of the blast injection type is to be driven from a crank on the main crank shaft running at 250 R.P.M. The compressor has to compress 1000 cu. ft. of free air per hour from atmospheric pressure (14·7 lb. per sq. in.) to 1000 lb. per sq. in., and the mean piston speed is limited to 400 ft. per min. Neglecting clearance, obtain principal dimensions for the compressor cylinders.

7. In an air compressor air is taken in at 20 lb. per sq. in. and 15° C., is compressed adiabatically to 100 lb. per sq. in. and is then expelled from the cylinder at that pressure. Find how much work is done on the compressor per lb. of air.

The same compression, from 20 to 100 lb. per sq. in., is then carried out in a two-stage compressor, the air, at pressure p lb. per sq. in., cooling down to 15° C. between the two cylinders. Find how much work is saved, per lb. of air, by the two-stage process, taking p as 30, 40, ... 90 successively, and plot a curve to determine the value of p which gives the greatest saving of work. Also shew that theoretically this value of p is 44·7 lb. per sq. in.

If the compression is carried out in two stages, 14·7 to 44 lb. and 44 to 120 lb. respectively, the air cooling to atmospheric temperature when passing between the compressing cylinders, the expansion being as before in a single-motor cylinder, find the efficiency.

8. If the clearance volume in a cylinder of a compressor is x times the piston displacement volume V, and if the pressure ratio of compression is r, prove that the volume of gas taken in per stroke when measured at the induction pressure is

$$V(1+x-xr^{\frac{1}{\gamma}}),$$

assuming adiabatic expansion and compression.

A single-stage compressor compresses air from 15 to 150 lb. per sq. in., the clearance volume being 1 per cent. of the piston displacement volume. At what point of the suction stroke will the induction valve open if the spring on it has negligible tension?

PAPER XXXIV

1. In a Bell-Coleman machine air is drawn into the compressor at atmospheric pressure and at a temperature of 15° C., and is compressed adiabatically to $4\frac{1}{2}$ atmospheres. It enters the expansion cylinder at 4 atmospheres pressure and at a temperature of 18° C., and the expansion is adiabatic. If the work put in the compressor is 10 per cent. greater than the indicated, and the work got out of the expansion cylinder is 10 per cent. less than the indicated, estimate the refrigerating effect per min. per each indicated horse power developed in the steam cylinder—both compressor and expansion cylinder being supposed without clearance. What results would you expect to get in practice with a machine working between the above limits of temperature?

2. A refrigerating machine using air and working on a reversed Stirling cycle maintains a temperature of −30° C. in the cold coils, the high temperature being 15° C. The ratio of expansion is 2 to 1. What is the coefficient of performance (1) when it is fitted with a perfect regenerator, (2) when the regenerator has an efficiency of 50 per cent.?

3. Describe the working of an air refrigerating machine in which the reversed Joule cycle is used, and sketch the p-V and ϕ-T diagrams. Find the coefficient of performance, and the highest temperature reached, if the air expands from a tempera-

ture of 16° C. to a temperature of −55° C. and has a temperature of −5° C. at the beginning of compression.

Also find the coefficient of performance if the air after leaving the cold chamber at a temperature of −5° C. has its temperature raised to 10° C. before compression begins, the temperature and pressure of the receiver of heat remaining the same.

4. A refrigerating machine using air and working on a reversed Stirling cycle works between temperatures 20° C. and −40° C. The ratio of expansion is 2·5 to 1. What is the coefficient of performance (1) when the regenerator has its efficiency equal to unity, (2) when the regenerator efficiency is 0·55?

Determine also the values of the coefficient, for the same cases, when the limits of temperature are 15° C. and −15° C.

5. In a refrigerator, in which 1 lb. of air is the working substance and the cycle is a reversed Joule's cycle, draw the pressure-volume and the entropy-temperature diagrams when the higher pressure is 4 atmospheres, and the temperature before compression in the pump, and at the beginning of expansion in the motor, is 15° C. Determine the coefficient of performance.

6. A Bell-Coleman refrigerating machine works between a pressure of 1 atmosphere and a pressure of n atmospheres. If α be the theoretical coefficient of performance, shew that

$$\frac{1}{\alpha}=n^{\frac{\gamma-1}{\gamma}}-1.$$

If the high pressure be $4\frac{1}{2}$ atmospheres and if the temperature at the beginning of compression be 10° C., and at the beginning of expansion 18° C., find the quantity of heat absorbed, per lb. of air, in the cooling chamber.

7. An air refrigerator working upon the Joule constant pressure cycle draws air into the compressor at a pressure of 15 lb. per sq. in. and a temperature 0° C. and compresses it to 60 lb. per sq. in. It is afterwards cooled by circulating water to a temperature of 20° C. and then expanded. The air expands and is compressed adiabatically according to the law $pV^{1\cdot4}=$constant.

Determine the horse power of the compressor and the volume of the compressor cylinder in order to produce 5 lb. of ice per min. from water at 10° C. The latent heat of water is 80 TH.U. per lb. and the volume of 1 lb. of air is 12·39 standard cu. ft. The compressor is single acting and runs at 100 R.P.M.

8. Describe the working of an air refrigerating machine in which the reversed Joule cycle is used, and sketch the p-V and ϕ-T diagrams for the cycle.

The cooling chamber is at a pressure of 15 lb. per sq. in. and is maintained at 0° C. 100 gallons of water passing through a coil of pipe in the chamber per hour are cooled through 10° C. If the ratio of compression in the compressor is 1·5, find the horse power required to drive the compressor, neglecting mechanical losses and heat leakage.

PAPER XXXV

1. An ammonia refrigerating machine is required to produce 3 tons of ice per hour from water at 12° C. The temperatures of evaporation and condensation of the ammonia are $-10°$ C. and 25° C. respectively, and the condensed ammonia is driven through an expansion valve without preliminary cooling. Find the horse power required in the compressor if the vapour is compressed adiabatically and is just dry at the end of compression. The latent heat of water may be taken as 80.

If the vapour is just dry at the beginning of compression and is compressed adiabatically to the same final pressure, find the amount of superheat at the end of compression and the horse power required.

2. A CO_2 machine working continuously produces $\frac{1}{2}$ a ton of ice per hour at a temperature of $-5°$ C. from water at 10° C. The temperature in the evaporator is $-10°$ C. and the pressure in the condenser is 1200 lb. per sq. in., the fluid leaving it at 20° C.

Assuming evaporation to be just complete and the compression therefore dry, find the quantity of CO_2 circulating in the machine per hour and the net horse power expended in compression.

[Latent heat of ice = 80, specific heat = 0·5.]

Under the same conditions of working, a vapour-compression refrigerating machine with throttle expansion gives a higher coefficient of performance when NH_3 is used as the working substance than when CO_2 is used. Explain this.

3. In the case of a machine using ammonia as working substance, determine the number of units of heat absorbed at the lower temperature per lb. and the volume of the compressor cylinder per thousand units of heat absorbed, when the limits of temperature are 20° and −30° C., the vapour in both cases being dry saturated at the end of compression.

Estimate the percentage increase in the coefficient of performance if, keeping the liquid pressure the same, the liquid is given a further cooling down to 10° C. before passing through the throttle valve.

4. A vapour refrigerating machine works with expansion through a nozzle, and the vapour is just dry at the end of the adiabatic compression. The limits of temperature are 15° C. and −25° C. Tabulate the following quantities for NH_3 and for SO_2:

(1) The dryness fraction at the lower temperature (*a*) at the beginning of evaporation, (*b*) at the beginning of compression.

(2) The heat absorbed, per lb. of the substance, at the lower temperature.

(3) The amount of substance required to absorb 1000 TH.U.

(4) The volume of cylinder swept through in the compressor per 1000 TH.U. absorbed.

(5) The work done in the compressor (*a*) per lb. of substance, (*b*) per 1000 TH.U. absorbed.

(6) The coefficient of performance.

5. An ammonia refrigerating plant is used for cooling a sterilised liquid having a specific heat of 0·96.

The following data represent mean values of the readings taken at 10-minute intervals:

Inlet temperature of the liquid 23·8° C.

Outlet temperature of the liquid 12·3° C.

Weight of liquid cooled per minute 156·5 lb.

Horse power exerted on the compressor shaft 15·3.

Inlet temperature of the cooling water at the ammonia condenser 8·9° C.

Outlet temperature of the cooling water 20° C.

Cooling water per minute 153 lb.

Heat rejected to compressor jacket per minute 83·5 TH.U.

Cooling effect on bare pipes per minute 222 TH.U.

Make out a balance sheet for this plant and determine the TH.U. not accounted for, and the ratio of the refrigerating effect to the work done in producing it.

6. In a simple CO_2 vapour-compression refrigerating plant with nozzle expansion, the absolute pressures on the suction and delivery sides of the compressor were 360 and 1200 lb. per sq. in. respectively. After leaving the condenser, the liquid CO_2 was given a further cooling down to 16·5° C. before it reached the regulator. Assuming that the CO_2 entered the compressor in the dry saturated condition, find by means of the I-ϕ chart the theoretical values of (1) the coefficient of performance and (2) the I.H.P. required per kilogram calorie refrigerating effect per sec. [N.B. 1 kilogram = 2·205 lb.]

The actually measured refrigerating effect per hour was 57,800 C.H. units, with 13·15 I.H.P. at the compressor. Find the actual values of (1) and (2).

7. A refrigerating machine, using ammonia as the working substance, makes 16 cwt. of ice per hour from water at 10° C. The temperature in the refrigerator is −10° C. and the pressure at the end of compression is 170·8 lb. per sq. in. The vapour is condensed at constant pressure and the liquid is cooled to 20° C. after which it flows through an expansion valve into the refrigerator.

If 9·6 lb. of ammonia pass through the refrigerator per minute, find the ideal coefficient of performance, the theoretical horse power required and the state of the ammonia at the end of compression. The latent heat of ice may be taken as 80.

8. Why is it desirable to have a large ratio between the latent heat of the vapour and the specific heat of the liquid in the fluid used in a vapour-compression refrigerator?

In the test of a single-acting ammonia machine, the brine tank was electrically heated and the details were as follows:

Temperature of brine tank ...	−5° C.
Condenser temperature ...	15° C.
Electrical input to brine tank	3650 watts
Flow of ammonia	0·86 lb. per min.
I.H.P. of compressor	0·96
Compressor cylinder... ...	4 in. stroke, 3·5 in. bore
R.P.M.	216
Volumetric efficiency ...	80 per cent.

Estimate:

(*a*) the coefficient of performance;

(*b*) the dryness of the ammonia at the beginning of compression;

(*c*) the theoretical coefficient of performance for the same temperature limits, the dryness at the beginning of compression being 0·78.

The compressor clearance volume may be neglected.

PAPER XXXVI

1. Determine the area of the throat of each of the four nozzles of a 30 H.P. De Laval turbine using 30 lb. of steam per H.P. hour. The steam enters the nozzle dry and at a pressure of 150 lb. per sq. in. and the expansion of the steam is according to the law $pV^{1\cdot3}=\text{constant}$.

2. 25 lb. of saturated steam per minute escape through a nozzle from a boiler at 150 lb. per sq. in. pressure: at the mouth of the nozzle where its sectional area is 1 sq. in., the pressure of the steam is 15 lb. per sq. in. Prove that the velocity of the steam at this point is about 1750 ft. per sec., and find its temperature. The specific heat of superheated steam at constant pressure may be taken as 0·5.

3. Steam initially dry and at rest under 180 lb. pressure flows down a nozzle, with no external communication of heat. It is estimated that at the section where the pressure is 40 lb. per sq. in. the kinetic energy of the steam is 20 per cent. less than it would be at that pressure in the case of frictionless flow. Find the percentage increase of section, at this pressure, as compared with frictionless flow, in order that the discharge may be the same in both cases.

Also if the pV curve during the expansion down the nozzle is of the form $pV^n = \text{a constant}$, find the value of n, and determine how much heat has been communicated, to each pound of steam, between the given pressures.

4. A well-designed nozzle takes dry saturated steam at 115 lb. per sq. in. and expands it to 4 lb. per sq. in. The final diameter of the nozzle is $\frac{1}{2}$ in. In the process 10 per cent. of the heat drop is not turned into kinetic energy. Find the velocity and state of the steam at issue from the nozzle and the weight of steam discharged per second.

5. A boiler maintains a supply of dry steam at a pressure of 100 lb. per sq. in., and the steam is blown off into the atmosphere through a converging-diverging nozzle, the diameter of the nozzle at its narrowest point being $\frac{1}{2}$ in. Assuming that the flow is adiabatic and frictionless, find how many lbs. of steam are discharged per hour.

6. In a test on a steam nozzle, the steam supply was at a pressure of 22·49 lb. per sq. in., whilst the temperature corresponded to a volume per lb. of 20·68 cu. ft. The discharge pressure was 14·77 lb. per sq. in. Assuming that the superheated steam follows the law $pV^{1\cdot3} = \text{const.}$ in adiabatic expansion, calculate the theoretical velocity of discharge. If the actually measured velocity was 1248 ft. per sec., what was the velocity coefficient of the nozzle?

7. Assuming that the steam is in thermal equilibrium throughout, design a suitable nozzle to supply 5 lb. of steam per minute, the steam being supplied (dry saturated) at $p = 120$ lb. per sq. in. and discharging at 2 lb. per sq. in. The critical pressure may be taken as $0\cdot58p$, whilst 12 per cent. of the total available heat drop is to be regarded as lost in friction in the divergent portion of the nozzle.

8. Steam at a pressure of 150 lb. per sq. in. and temperature 250° C. expands through a ring of nozzles to a pressure of 20 lb. per sq. in. The kinetic energy of the issuing steam is found to be 0·94 of the energy which it would have had if it had expanded adiabatically and in thermal equilibrium. Find the discharge area of the nozzles for an outflow of 2500 lb. of steam per minute.

PAPER XXXVII

1. Steam expands adiabatically from 100 lb. per sq. in. dry saturated to 20 lb. per sq. in. in a nozzle inclined at 10° to the plane of an impulse turbine wheel. If the mean blade velocity is 1500 ft. per second, determine the inlet angle of the blade.

Find also the work done per lb. of steam if the outlet angle of the blade is 45°.

Friction may be neglected.

2. In a turbine of De Laval type steam is supplied at 150 lb. per sq. in., dry saturated, and is expanded to 2 lb. per sq. in., the jets are set at an angle of 20°, and the blade speed is 1200 ft. per sec. If 90 per cent. of the frictionless adiabatic heat drop is converted into kinetic energy and the axial velocity of flow is constant, find the correct entry and exit angles of the blades and the ideal efficiency of conversion of kinetic energy into work.

Explain briefly how in turbines the blade speed can be reduced without loss of efficiency.

3. In a De Laval turbine, in which the full pressure drop is used up in the nozzles and the blades have their inlet and outlet angles equal, steam is supplied dry and saturated at 180 lb. per sq. in. and the exhaust pressure is 2·5 lb. per sq. in. The peripheral speed of the blades is 1250 ft. per sec. and the nozzles make an angle of 20° with the direction of motion of the blades. Assuming adiabatic flow in the nozzles and neglecting friction, find the velocity of discharge from the nozzles, the inlet and outlet angles of the blades and the work done per second per pound of steam.

4. A De Laval turbine has blade angles of 30° at inlet and 38·5° at outlet. The nozzle axis makes 20° with the plane of the disc. The steam enters the blades without shock and leaves the blades with a relative speed which is 80 per cent. of the relative speed at entry. The steam is dry and saturated at entry and expands from 100 to 1 lb. per sq. in. in the nozzle. Determine the thermal efficiency of the turbine, neglecting the losses in the nozzle; the temperature of the boiler feed being 38° C.

5. At one stage of the Curtis turbine the exit velocity from the nozzles is 1600 ft. per sec., the mean blade ring diameter is 40 in., and the speed of the turbine 3000 R.P.M. The angle of the nozzle is 20°, and of the blade at exit 25°. Find the work done on the blades per lb. of steam. Assume that the velocity in the blade passages relative to the wheel at exit is 0·8 of the velocity at entrance. Determine also the axial thrust of the steam per lb. and the energy carried to the next stage.

6. The figure shews, diagrammatically, the high-pressure blading of a Parsons turbine. The blade speed is 200 ft. per sec. Shew that if all losses are neglected, 62 rows of blades (half fixed and half moving) will be required to drop the pressure from

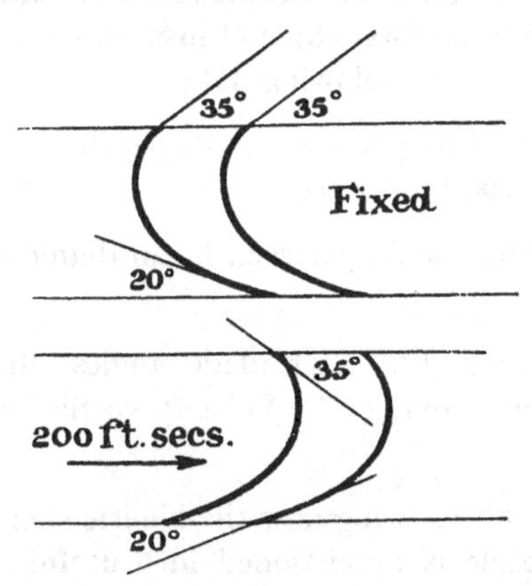

200 lb. per sq. in. to 25 lb. per sq. in., the steam being dry and saturated at entry and the blade heights being graded so as to give a constant steam velocity. Assuming that as the effect of steam friction and other losses the efficiency of the turbine

is 60 per cent. of that of the Rankine engine, calculate the height of the blades at entry for a turbine giving 5000 H.P. at 750 R.P.M.

The higher the blade speed, the smaller is the number of rows, and the shorter the turbine, required to utilise a given drop of pressure. Explain why the blade speeds in marine turbines are, nevertheless, usually lower than in land turbines, though considerations of space are more important.

7. In a Parsons' reaction turbine the moving blades are similar to the fixed ones and the axial velocity of flow of the steam is constant. The blades have an outlet angle of 20° and the velocity of discharge from the moving blades is 120 ft. per sec. at an angle of 110° to the direction of motion of the blades. Draw the velocity diagram for a stage comprising one ring of fixed and one ring of moving blades, and determine (i) the blade velocity, (ii) the work done in the stage per lb. of steam, (iii) the efficiency of the stage if the mean volume of the steam there is 29·5 cu. ft. per lb. and the drop in pressure in the stage is 1 lb. per sq. in.

8. If the wheel in a pressure stage of an impulse turbine has two velocity stages, find the height of the second row of moving blades, assuming the final channel just runs full over the whole circumference, for the following data:

Steam used: 4 lb. per sec., cu. ft. per lb. 27·5. Exit velocity from nozzles: 2200 ft. per sec.

Blade speed: 450 ft. per sec., mean diameter of blade ring 15 in.

Nozzle angle: 20°. Exit blade angles: first moving 23°, fixed 23°, second moving 35°. Velocity coefficient in each blade passage, 0·85.

Make up a table shewing how the kinetic energy of the steam leaving the nozzle is apportioned into useful work, frictional losses in the passages, and energy carried over to the next stage.

PAPER XXXVIII

1. In a reaction turbine of the Parsons type, both the fixed and the moving blades have inclinations, at the receiving and discharging tips, of 35° and 20° respectively (in opposite directions) to the directions of motion of the blades. The mean diameter of the blade ring circle for a number of pairs of rows of fixed and moving blades is 2 ft. 7 in. and the speed of rotation is 1500 R.P.M.

Calculate the work developed in *six* consecutive moving rows of blades per lb. of steam, assuming no shock at entry to the blades. If the steam leaves the *sixth* pair at 20 lb. per sq. in. and dryness 0·91, and the efficiency ratio for the six pairs be 80 per cent., find the total actual and theoretical heat drops for the six pairs and the state of the supply steam to the first of the six pairs of blades. Graphical methods may be used.

2. If I_1, I_x, I_2 are the total heats of the steam under the conditions of supply, bleeding and exhaust, and I_{w2} the total heat of the condensed steam, shew that the efficiency of the cycle is

$$\frac{I_1 - mI_x - (1-m)\,I_2}{I_1 - mI_x - (1-m)\,I_{w2}},$$

and that bleeding increases the efficiency at whatever stage the steam is tapped off.

Steam is supplied to a turbine at a pressure of 200 lb. per sq. in. and temperature 300° C., and a pressure of 0·8 lb. per sq. in. is maintained in the condenser. What proportion of the steam should be tapped off at a pressure of 25 lb. per sq. in. in order to give the best results?

3. The moving blades in one stage of a Curtis turbine have a mean blade-ring diameter of 50 in., and the blades at exit make an angle of 25° with the plane of revolution. The turbine speed is 2500 R.P.M. If the angle of the delivery nozzles is 20° and the exit velocity of the steam from the moving blades 1500 ft. per sec. relative to them, find the work done on the moving blades per lb. of steam.

Assume that the velocity of the steam relative to the moving blades at exit is, on account of friction, 0·85 of the relative velocity at entrance.

What is the axial thrust per lb. of steam per sec. on the moving blades and the energy carried on to the next stage?

4. An ideal steam turbine plant is supplied with superheated steam at 550 lb. per sq. in. and temperature 400° C. ($I = 771{\cdot}7$, $\phi = 1{\cdot}63$.) If the feed water be supplied at 38·9° C., and if the whole of the frictionless adiabatic heat drop is utilised and the expansion is down to 1 lb. per sq. in., find the thermal efficiency and the steam supply per shaft H.P. hour.

Find the corresponding quantities when feed water heating by 'bled' steam is employed. The feed heating may be taken as ideal, i.e. with an infinite number of stages, and starting when the steam is just dry saturated. What total percentage of the steam supplied to the turbine is tapped off?

5. Shew that, theoretically, the operation known as *bleeding* in a steam power plant must increase the efficiency of the plant no matter what proportion of bleeding is applied (within certain limits) nor at what intermediate temperature the tapping is made.

Shew that between tapping pressures of 25 and 50 lb. per sq. in. there is very little change in increase of efficiency obtainable by bleeding in a plant in which the supply steam is dry saturated at 200 lb. per sq. in. and the condenser pressure is 2 lb. per sq. in.

6. The figure shews consecutive rings of fixed and moving blades of a reaction turbine. All the blades are of the same shape, their radial length being increased to accommodate the increasing volume of the steam as it flows through the turbine.

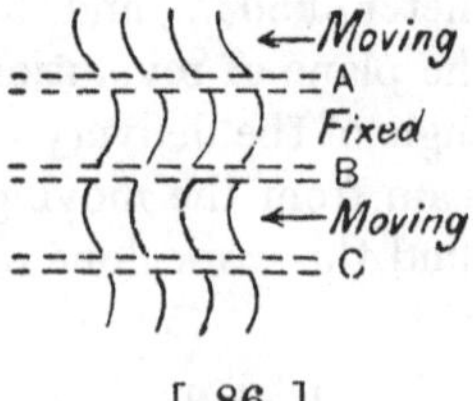

If the velocity of the steam entering the fixed blades at A is equal to that entering the next set of fixed blades at C, and is parallel to the blades, prove

(1) that the heat drop in each row of blades, fixed or moving, is the same, namely $I_A - I_B = I_B - I_C$;

(2) that the work done on each set of moving blades is $2(I_B - I_C)$.

What is 'end-tightening' as applied to reaction blading and how has its introduction affected the relative position of the impulse and reaction types?

7. Explain why the process of bleeding increases the efficiency of a steam turbine.

A turbine is supplied with steam at 200 lb. per sq. in. and 350° C.; the condenser pressure is 0·4 lb. per sq. in.; and sufficient steam is bled at 20 lb. per sq. in. to heat the feed water to 100° C. Calculate the percentage of the total steam consumption which must be tapped off

(*a*) if the expansion is frictionless and adiabatic;

(*b*) if the expansion is such that for the two parts of the turbine, before and after the tapping point, the ratio

$$\frac{\text{work done on rotor}}{\text{ideal heat drop}}$$

is equal to 0·75.

Compare the two efficiencies.

8. Shew that constant pressure lines in the wet region of an I-ϕ diagram for water-steam are straight and of slope represented by the absolute temperature corresponding to the pressure.

In a compound impulse turbine the intermediate pressures are arranged to correspond to equal adiabatic heat drops in each stage. The steam expands from 225 lb. per sq. in., 300° C., to 80 lb. per sq. in. in the first three stages. Estimate the intermediate pressures.

If the stage efficiency is found to be 75 per cent. for each stage, find the state of the steam leaving the third stage.

Give a sketch shewing how you have obtained your results.

Carry-over and radiation losses are to be neglected.

PAPER XXXIX

1. A cold-storage chamber $8 \times 10 \times 12$ ft. in size is protected with non-conducting material of thickness 6 in. The coefficient of conductivity of the material is, in C.G.S. units, 0·0003. A temperature of $-5°$ C. is to be maintained in the chamber, with an outside air temperature of $+25°$ C., by some form of refrigerator.

How many C.H.U. per minute will have to be removed by the refrigerator?

2. A surface condenser has 1·8 sq. ft. of surface per I.H.P. and the engine uses 20 lb. of water per I.H.P. per hr. If the heat rejected to the condenser is 600 TH.U. per lb. of steam, find the number of TH.U. conducted away per sq. ft. per min. Moreover, assuming that 0·4 TH.U. is conducted away per sq. ft. per sec. per degree difference of temperature between the steam and the plate, estimate the drop in temperature between the steam and the plate.

3. If in cast-iron the flow of heat be 0·175 gramme-calorie per sec. across a square centimetre when the temperature gradient is 1° C. per centimetre, find the value of the flow in B.TH.U. per sec. across 1 sq. ft. when the temperature gradient is 1° F. per ft.

Also, if there is a temperature gradient of 12° C. per cm. in the same metal, find the rate of flow of heat, in horse power, across 1 sq. ft.

4. A steam main 6 in. in external diameter conveys 120 lb. of steam per min. from a boiler working at 200 lb. pressure per sq. in. to an engine 90 ft. away. The pressure at the engine stop valve is 150 lb. and the dryness fraction of the steam is 0·93. Calculate the loss per min. in TH.U. per sq. ft. of pipe due to conduction and radiation, if the effective length of the pipe is 100 ft. The pipe is afterwards covered by lagging and the steam then reaches the engine at a pressure of 190 lb. per sq. in. and dryness fraction 0·97. Calculate the saving in TH.U. per min. effected by the lagging.

5. Assuming that the rate of conduction of heat from the hot gases to the water in a boiler is proportional to the square of the difference in temperature between the two, and that the air supply per lb. of coal is sensibly constant under all loads, shew that the relationship between the weight of water evaporated from and at 100° C. per hour, and the weight of water evaporated from and at 100° C. per lb. of coal consumed is, very approximately, represented by a straight line law when the boiler is tested under different rates of combustion.

6. The coefficient of transmission through the tubes of a condenser is 1000 gramme-degree units of heat per hour per sq. metre of surface for each degree centigrade of temperature difference between the outside and inside of the tube. Express the rate of flow in horse power per sq. ft. when the difference of temperature is 25° C.

In a certain condenser the value of the coefficient in the units given above is K. Water enters a tube at temperature θ_0 and leaves it at temperature θ_1, and the tube is surrounded by steam at temperature t. The area of the surface of the tube is A sq. metres. If W grammes of water pass per hour, shew that

$$W \log_e \frac{t-\theta_0}{t-\theta_1} = K.A.$$

7. A constant temperature difference of θ is maintained between the inner and outer surfaces of a long cylindrical tube, which has internal radius r_1, external radius r_2, and conductivity k. Shew that the flow of heat through the walls of the tube, per unit length, is

$$2\pi k \frac{\theta}{\log_e \dfrac{r_2}{r_1}}.$$

For a series of tubes, ranging from 2 in. to 20 in. in internal diameter, and each $\frac{1}{2}$ in. thick, the flow of heat is calculated on the assumption that there is a uniform temperature gradient in the walls. Plot a curve shewing how the percentage error made varies with the diameter.

8. The radii of the internal and external surfaces of a circular cylinder are r_1 and r_2 and these surfaces are kept at temperatures

t_1 and t_2. If k be the conductivity of the material, prove that in the steady state the quantity of heat which flows per sec. across a length l of the cylinder is

$$\frac{2\pi kl\,(t_1-t_2)}{\log r_2-\log r_1}.$$

PAPER XL

1. A steel rod has its ends embedded in massive blocks. It measures 10 cm. long between the blocks and its diameter is 6 mm. The temperature of one block is 20° C. and of the other 10° C., and the rod is heated by a steady electric current flowing along it, the total energy dissipated in it being 5 watts. Assuming that the rod does not lose heat except by conduction into the blocks, find the steady distribution of temperature along it. The conductivity of the steel may be taken as 0·15.

2. In an experiment to determine the rate of condensation of steam on a clean metal surface, the apparatus employed consisted of a cast-iron cylinder having an internal diameter of 1 in. and an external diameter of 5 in., with two mercury thermometers let into mercury pockets at distances of $1\frac{1}{2}$ in. and 2 in. from the axis. The outer surface was kept cool by rapidly circulating water, and the steam was condensed on the inner face of the cylinder. When the condensation amounted to 15·9 TH.U. per sq. ft. per sec., with a steam temperature of 135° C. the observed temperatures at the pockets were 57° and 40° C. respectively. Deduce the inner surface temperature of the metal and the condensation in TH.U. per sq. ft. per sec. per degree difference of temperature between the steam and the metal.

3. In some experiments on cooling from bare and covered steam pipes, the following figures were obtained:

Loss of heat per hr. per sq. ft. of bare pipe ... =770 TH.U.

Loss of heat per hr. per sq. ft. of covered surface with lagging 1 in. thick ... ... =100 TH.U.

Loss of heat per hr. per sq. ft. of covered surface with lagging $2\frac{1}{2}$ in. thick ... ... = 53 TH.U.

If 120 lb. of steam per hour pass through a 4 in. pipe 30 ft. long, steam being 226 lb. per sq. in. and 200° C., find the dryness fraction of the steam on leaving

(1) the bare pipe,

(2) the pipe with 1 in. thick covering,

(3) the pipe with $2\frac{1}{2}$ in. thick covering.

4. The internal and external surfaces of a sphere are kept at temperatures θ_1 and θ_2. Shew that entropy is generated in the sphere at the rate of $\frac{4\pi k r_1 r_2}{r_2 - r_1} \frac{(\theta_1 - \theta_2)^2}{\theta_1 \theta_2}$, where r_1, r_2 are the internal and external radii of the sphere.

5. Water at temperature θ_1 passes into a tube of length l and diameter d, and leaves the tube with a temperature θ_2; the temperature of the tube itself being θ_0. The heat given by the tube to the water is $\alpha \rho w (\theta_0 - \theta)$, where α is a constant, ρ = the density of the water, w = its velocity, and θ = its temperature. Shew that the rise of temperature is given by

$$\theta_2 - \theta_1 = (\theta_0 - \theta_1)(1 - e^{-\mu}),$$

where $\mu = \frac{4\alpha l}{\sigma d}$, σ being the specific heat of the water.

6. In a condenser the heat abstracted from the steam per unit area of tube surface at any point is $cv(t_1 - t)$, where t_1 is the constant temperature of the steam, t that of the water, and v its velocity. Prove that the rise of temperature of the circulating water is $(t_1 - t_0)(1 - e^{-\frac{4c}{\rho\sigma} \cdot \frac{l}{d}})$, where l is the length of the tubes, d their diameter, ρ the weight of a cu. ft. of water, and σ its specific heat.

7. Shew that if the fluctuation of temperature at the surface of a gas-engine cylinder wall is $\pm T$ and is harmonic, the temperature at any depth x in the cylinder wall at any time t is given by $Te^{-\alpha x} \cos(\theta - \alpha x)$, where $\alpha = \sqrt{\frac{\pi N}{K}}$, N is the number of revolutions per unit time, θ is $2\pi N t$ and K is the diffusivity of the metal.

The temperature fluctuation at the wall surface of a cylinder was found to be $\pm 20°$ C. during a cycle. Determine the depth at which this is reduced to 1° C., and the fluctuation when the temperature changes are in step with those at the surface, if $N=240$ per min. and $K=1\cdot2$ (inch, minute and Centigrade degree units).

8. An infinite block of homogeneous material with a plane face, its average temperature being θ_0, has a layer of water on the face of weight w unit area. The water is exposed to steam whose temperature alters with time according to the law $\theta=\theta_0+\theta_1 \sin(qt+c)$. The water is kept at the temperature of the steam by condensation and vaporisation, but receives no increase or diminution except in this way. The emissivity between water and metal is e. Shew that there is a diminution of water at the end of every cycle which does not depend upon the amount of water but only upon e and the character of the metal.

PAPER XLI

1. A gas expands so that $pv^n=$constant, and the external work done is supplied half from the internal energy of the gas and half from an outside heat supply; shew that $n=\frac{\gamma+1}{2}$, where γ is the ratio of the specific heats.

2. A torpedo air chamber contains initially 80 lb. of air at a pressure of 1700 lb. per sq. in. and 15° C., and at the end of the run the pressure is 500 lb. per sq. in. and the temperature 2° C. How much of the heat in the air which is left in the chamber has been abstracted from the sea?

3. An engine receives 150 TH.U. during a rise of temperature from 100° to 200° C. at a rate which is proportional to the rise of temperature, also 500 TH.U. at the constant temperature of 200° and 50 TH.U. more whilst the temperature falls from 200° to 100°, the rate of supply being then proportional to the change of temperature. 100 TH.U. are rejected at the temperature of 200° and the remainder which is not utilised is rejected at a tempera-

ture of 100°. Estimate the greatest efficiency of such an engine and compare it with that of a reversible engine which works between the limits of temperature of 200° and 100° C.

4. Assuming the relation between the pressure and volume of dry saturated steam to be represented by the equation

$$(p+\alpha)\,v=\beta,$$

where α, β are coefficients, shew that when the horse power of a given engine is varied by varying the boiler pressure—the cut-off, speed and back pressure remaining unaltered—the relation between the weight of steam used per hour and the indicated horse power is represented by a linear law. Clearance and all losses may be neglected, the steam assumed dry at cut-off, and the curve of expansion to follow the law

$$pu^n=\text{const.},$$

where u is the volume of the mixture.

On what experiments on steam-engine performance has this question a bearing?

5. A steam turbine plant containing H.P. and L.P. portions is supplied with superheated steam at 550 lb. per sq. in. and 400° C. ($I=771{\cdot}7$). The steam leaves the H.P. portion superheated at 90 lb. per sq. in. and 195° C. ($I=679{\cdot}6$) and then passes through a heater where (owing to friction) the pressure falls to 80 lb. per sq. in. but the temperature is raised to 360° C. In this condition ($I=763{\cdot}7$, $\phi=1{\cdot}825$) the steam enters the L.P. portion and expands down to 1 lb. per sq. in.; the work in this portion is 85 per cent. of the work that could be obtained with frictionless adiabatic expansion. Find per lb. of steam: (1) the work done in the H.P. portion, (2) the heat supplied in the heater, (3) the work done in the L.P. portion. Neglect external losses in the turbine, and the kinetic energy of the steam in the heater.

If the feed water be supplied at 38·8° C., find the overall thermal efficiency of the plant, and the steam supply per shaft H.P. hour.

6. The steam supply to an engine with jacketed cylinders is at a pressure of 100 lb. per sq. in. and 0·9 dryness. Similar steam is used for the jackets and may be assumed to be just

condensed there and then drained away without any reduction in temperature. The ratio of expansion for the engine is 14 and expansion follows the law $pV^{1.05}$=constant. The condenser pressure is 2 lb. per sq. in.; clearance and compression may be neglected.

Shew that, per lb. of steam used in the cylinder, the work done is approximately the same as in a Rankine engine using similar steam between the same limits of pressure.

Compare the thermal efficiencies of the two engines, if 0·2 lb. of steam is condensed in the jackets per lb. of steam used in the cylinder.

7. Shew that, in the steady flow of a fluid through a heat plant of any description, the change of the *I* of the fluid, in any period, represents the sum of the heat conversion into external work and heat reception by the plant during the same period.

Dry air at a temperature of 20° C. is blown through the tubes of a condenser along with the circulating water, which enters at 10° C. The air and water issue from the condenser at 15 lb. per sq. in. and 35° C., the air being fully saturated with water vapour. If 5 lb. of air are supplied per lb. of circulating water, determine the weight of circulating water required to extract 1000 C.TH.U. from the condenser.

8. The analysis of a sample of dry exhaust gas from an engine running on a rich mixture of petrol C_xH_y and air is (by volume) as follows: CO_2=8·20 per cent., CO=9·74 per cent., O_2=Nil, H_2=3·47 per cent., CH_4=1·14 per cent., N_2=77·45 per cent. Shew that the ratio of air to petrol by weight is about 10·5 : 1, and find the ratio of y to x.

[H_2=2, O_2=32, N_2=28; composition of air by volume is N_2=79 per cent., O_2=21 per cent.]

9. A gas engine has a clearance volume of 1 cu. ft. and a ratio of compression 6. The volumetric efficiency is 75 per cent. when the prevailing engine-room conditions are 15° C. and 14·7 lb. per sq. in. At the beginning of the suction stroke the clearance volume contains residual gas at 700° C. and 14·7 lb. per sq. in. If the pressure remains constant throughout the suction stroke,

estimate the temperature of the cylinder contents at the end of the suction stroke and the heat received from the walls, valves and passages.

It may be assumed that all the gases have the same specific heats as air.

10. A CO_2 refrigerating machine working with single-stage compression gave the following results on test:

Heat rejected to cooling water per min.	387 C.H.U.
Heat removed per min. by the brine from the cold room	238 C.H.U.
Total I.H.P. of compressor	4·73

Draw up a heat balance, and give the actual coefficient of performance.

If the absolute pressures in the condenser and evaporator are respectively 965 and 334 lb. per sq. in., and the vapour is just dry saturated at the end of compression, find the theoretical coefficient of performance.

PAPER XLII

1. A deflated gas bag of small capacity is charged from a steel bottle containing gas under a high pressure and at a temperature of 15° C. When the valve between the bag and the bottle is closed, the pressure in the bag is 0·5 lb. per sq. in. by gauge. Assuming that the fabric of the bag does not stretch, and that there is no loss of heat during the process of charging, shew that the temperature of the gas in the bag is about 17·7° C. The height of the barometer is 29 in. and γ for the gas is 1·4.

2. A quantity of heat Q units is being generated at 100° C. and is available for heating a room, but a perfect heat engine is set to take the heat Q and drive a perfect reversible engine which takes heat from the atmosphere at 15° C. and delivers heat to the room at 30° C. Shew that the maximum quantity of heat that can be delivered to the room is 4·6 Q units. The lowest available temperature is that of the atmosphere.

3. Two points on the expansion line of an indicator card gave

(1) $p_1 = 75$ lb. per sq. in. and (2) $p_2 = 15$ lb. per sq. in.
$v_1 = 1{\cdot}3$ cu. ft. $v_2 = 5{\cdot}1$ cu. ft.

the pressures being measured from the atmospheric line and the volumes from the beginning of the stroke. Atmospheric pressure 15 lb. per sq. in.

The total mass of steam expanding in the cylinder was estimated at 0·48 lb. and the volume of the clearance space 0·46 cu. ft. Assuming the expansion line to be represented by the equation $pv^n = \text{constant}$, find the work done by the steam between the points (1) and (2) and the interchange of heat between the steam and the cylinder.

4. State Dalton's law of partial pressures for mixtures of different gases, and shew how the law can be explained by the molecular theory.

The condenser of a turbine is to be designed to maintain a vacuum of 27 in. of mercury. It is known that the steam in the turbine is liable to carry with it air amounting to 2·5 per cent. of its own weight. Assuming that there is no leakage in the air pump, calculate from the following data the necessary volume swept by the pump piston per horse power of the turbine per hour:

Steam used by turbine per H.P. hour 11·5 lb.

Temperature of air pump discharge 40° C.

Barometer 30 in.

5. A gas producer in which air and steam are drawn through incandescent anthracite is supplying gas consisting of hydrogen, carbon monoxide and nitrogen only.

Analysis of coal: carbon 94 per cent., ash 6 per cent. by weight.

Analysis of air: oxygen 23·2 per cent., nitrogen 76·8 per cent. by weight.

Heat of formation: steam 34,500 TH.U. per lb. of hydrogen, carbon monoxide 2450 TH.U. per lb. of carbon.

Assuming no heat interchanges other than those in the

chemical reactions, shew that 0·6 lb. of steam and 3·1 lb. of air will be required per lb. of coal burnt.

Determine the volumetric analysis of the producer gas.

6. In an engine working on the Otto cycle release occurs at 80 per cent. of the stroke and at a pressure of 45 lb. per sq. in. The temperature of the gas and air drawn in is 50° C. and an explosion occurs every two revolutions of the engine. If the compression ratio is 5, estimate the temperature of the charge filling the cylinder at the end of the suction stroke. State what assumptions you make in arriving at this estimate. Atmospheric pressure 15 lb. per sq. in.

7. A refrigerator using air as the working fluid consists of a turbo-compressor and an expansion turbine mounted on the same shaft. Air at 0° C. and 15 lb. per sq. in. enters the compressor section and is raised in pressure to 50 lb. per sq. in.: it is then cooled to 15° C., the pressure remaining unchanged, and passes through the expansion section, doing work and falling in pressure to 15 lb. per sq. in.: it then passes through the cold store, abstracting heat, and re-enters the compressor. During expansion the law $pV^{1\cdot 35}=\text{const.}$ and during compression $pV^{1\cdot 45}=\text{const.}$ may be taken to hold.

Calculate the change of entropy during compression and expansion and sketch the T-ϕ diagram for the cycle. Also determine the heat abstracted per lb. of air circulating.

8. Illustrate the value of using a high vacuum in a turbine condenser by finding, from a steam diagram or otherwise, the kinetic energy in ft.-lb. per lb. of steam after expanding down a properly shaped nozzle from a pressure of 150 lb. per sq. in. and temperature 200° C. to pressures of 15, 5 and 1 lb. per sq. in. respectively, assuming in each case thermal equilibrium and no frictional loss. What would be the velocity of the steam in the last case (*a*) if there were no friction, and (*b*) if owing to friction only 85 per cent. of the available heat energy were converted into kinetic energy?

9. A De Laval turbine uses dry saturated steam at a pressure of 120 lb. per sq. in. and exhausts at 3 lb. per sq. in. The actual

heat drop in the nozzle is 85 per cent. of the frictionless adiabatic drop and the nozzles are inclined at 15° to the plane of the wheel. If the relative velocity is reduced by 12·5 per cent. as the steam passes over the blades and the blades at exit are inclined at 55° to the plane of the wheel, determine the velocity of the blades so that the steam will leave them axially.

It may be assumed that there is no shock at entry.

10. If the relation between the internal energy and the absolute temperature (τ) of a gas for which $\frac{pv}{\tau}=R$ be

$$E=C\tau+\lambda\tau^2+\text{const.},$$

where C and λ are constants, show that the relation between pressure and volume in an adiabatic change will be

$$pv^\gamma . e^{\frac{2\lambda}{C}\frac{pv}{R}}=\text{const.},$$

where $\gamma=\frac{C+R}{C}$.

For the products of combustion in a gas engine, C may be taken as 19 and λ as ·005 in ft.-lb. per standard cu. ft. The compression ratio is 6. Find the maximum pressure and the efficiency in a cycle of the ordinary type in which there is no loss of heat, and in which the combustion is complete and instantaneous, when the heat-supply is 50,000 ft.-lb. per standard cu. ft. (lower value). The temperature and pressure at the end of the suction stroke may be taken as 100° C. and 14·7 lb. per sq. in. respectively, and the contraction in combustion may be neglected.

PAPER XLIII

1. Air is compressed adiabatically into a receiver of capacity V, to m times the density of the atmosphere. Shew that if p be equal to the atmospheric pressure, the work expended is

$$pV\left(\frac{m^\gamma-\gamma m}{\gamma-1}+1\right).$$

2. In an experiment for determining γ, the ratio between the specific heats of air at constant pressure and constant volume, dry air was pumped into a vessel until the pressure inside was p_1. The atmospheric pressure outside was P. A valve was then opened for a short time, during which the pressure in the vessel rapidly fell to P, and after the valve was again closed it rose finally to p_2.

Assuming that the initial and final temperatures of the air in the vessel are the same and that the expansion while the valve is open is adiabatic, shew that

$$\gamma = \frac{\log p_1 - \log P}{\log p_1 - \log p_2}.$$

3. Illustrate on a T-ϕ diagram for steam, approximately to scale, the following processes:

(*a*) Adiabatic expansion in an engine cylinder from 200 lb. per sq. in. and dryness fraction 0·95 to 8 lb. per sq. in. release pressure, followed by condensation at constant volume to 2 lb. per sq. in.

(*b*) Adiabatic expansion in a turbine accompanied by re-superheating:

initial pressure 200 lb. per sq. in., initial temperature 250° C.
first expansion to 30 lb. per sq. in., re-superheated to 200° C.
second expansion to 2 lb. per sq. in.

Mark on the diagram the state of the steam after each stage of each process and find the heat which would be rejected to the condenser in each case, if complete condensation follows at constant pressure.

On a site for which the supply of condensing water is limited, it is required to install plant to give maximum power irrespective of efficiency; determine whether scheme (*a*) or (*b*) should be adopted.

4. A boiler of 1200 cu. ft. capacity contains 40,000 lb. of steam and water at a pressure of 50 lb. per sq. in.: if the efficiency of the boiler is 65 per cent., calculate the weight of coal to be burnt to raise the pressure to 200 lb. per sq. in. The calorific value of the coal is 7500 TH.U.

5. Not infrequently, in marine work, the boiler steam at a pressure of 225 lb. per sq. in. is wire-drawn by means of a reducing valve, to a pressure of 180 lb. per sq. in. before admission to the cylinder. Assuming the dryness fraction on the boiler side of the valve to be 0·95, calculate the dryness fraction on the engine side of the valve, and also—assuming the steam to expand in a single cylinder with non-conducting walls and without clearance or drop—estimate the percentage loss of work due to wire-drawing when the condenser pressure is 2 lb. per sq. in.

6. 10 lb. of steam at a pressure of 200 lb. per sq. in. and 1 lb. of air at a pressure of 14·7 lb. per sq. in. and temperature 20° C. are confined in the same vessel but separated by a partition. Shew that if the partition breaks down the temperature after mixture will be 173° C., assuming that no heat enters the vessel from outside.

Determine the resulting pressure. The volume of 1 lb. of air at 14·7 lb. per sq. in. and 0° C. is 12·39 cu. ft.

7. Describe with sketches the usual arrangement of 'economiser' as fitted with a Babcock and Wilcox boiler. With a view to testing for leakage of air into the economiser the flue gases were analysed, during a trial, both before and after their passage through it. The results are given below, together with the other observations made. Calculate the air leakage per lb. of coal burned and also what fraction of the heat in the flue gases, measured from 0° C., is given up to the feed water as it passes through the economiser.

	Entering economiser	Leaving economiser
Flue gas (dry):		
Analysis by volume	N_2 79 per cent.	N_2 79 per cent.
	CO_2 10·5 ,,	CO_2 9·7 ,,
	O_2 10·5 ,,	O_2 11·3 ,,
Temperature	410° C.	
Feed water:		
Temperature	12° C.	117° C.
Quantity	9500 lb. per hour	9500 lb. per hour

The coal contained 85 per cent. of carbon by weight, and no free hydrogen. The consumption was 1000 lb. per hour.

Assume the mean specific heat (at constant pressure) of the flue gases at entry to economiser to be 0·25.

Nitrogen: Oxygen in air by volume = 79:21.

Atomic weights: N = 14, C = 12, O = 16.

8. Explain carefully why in actual practice the efficiency of a Diesel engine can be greater than that of one working on the Otto cycle.

A four-cycle Diesel engine has adiabatic compression ($\gamma = 1{\cdot}38$) from 100° C. and 14·7 lb. per sq. in. to 500 lb. per sq. in. Oil is then injected and burns at the constant pressure of 500 lb. per sq. in. until the temperature of the cylinder contents is 1800° C., after which the temperature remains constant until the injection of oil ceases. The indicator diagram is completed by an adiabatic and a constant volume line. 1 lb. of the oil used required 180 standard cu. ft. of air for complete combustion giving out 14×10^6 ft.-lb. of heat energy. In the above engine the air actually used in burning the oil may be taken as 70 per cent. of the cylinder contents at the beginning of compression. The specific heat at constant pressure of the cylinder contents is 33 ft.-lb. per standard cu. ft. Treating the cycle as ideal, find the pressure at the end of the isothermal process.

9. Calculate the coefficients of performance of two vapour compression refrigerators with throttle expansion working between −10° C. and +20° C., the one using ammonia and the other carbon dioxide, the vapour being dry and saturated at the end of compression.

Shew that the heat extracted per cu. ft. of swept compressor volume is about four times as great with carbon dioxide as with ammonia.

10. A destroyer of 14,000 shaft H.P., driven by a Parsons' turbine running at 720 R.P.M., uses 14 lb. of dry saturated steam per shaft H.P. hour, the boiler pressure being 200 lb. per sq. in. The velocity of the first row of moving blades is 100 ft. per sec., the steam velocity as it leaves the first guide blades is 300 ft. per sec., and the outlet angle of the blades is 20°. Calculate the approximate height of the first row of blades.

PAPER XLIV

1. A vessel is exhausted of air to a pressure of 12 lb. per sq. in., the pressure of the atmosphere being 15 lb. per sq. in. The temperature of the whole being that of the atmosphere (15° C.), a cock is opened and air allowed to rush in until the pressure is equalised. Assuming that no heat is lost to the walls of the vessel, find the rise of temperature of the air within it.

2. Two vessels A and B of equal volume are connected through a stopcock. Originally the stopcock is shut; whilst A is completely evacuated and B is filled with air at absolute temperature T and pressure p. Shew that if the stopcock be gradually opened, pressure equilibrium will be established when the pressure in each vessel is $p/2$, and that the absolute temperatures (supposed uniform) in A and B will then be about $1{\cdot}28T$ and $0{\cdot}82T$ respectively. It is to be assumed that there are no heat losses.

3. Determine, using the steam tables, the increase of entropy of 1 lb. of water as it is brought from 15° C. to steam at temperature 150° C. and of dryness 0·85.

1 lb. of dry saturated steam is confined in a vertical non-conducting cylinder under a movable piston which is loaded so that the pressure in the steam is 150 lb. per sq. in. abs. and the volume of the steam is 1 cu. ft. If the load on the piston is suddenly reduced to give a pressure of 50 lb. per sq. in. abs. when at rest, determine the dryness of the steam when equilibrium has been established, and the change of entropy. Neglect the dissipation of energy due to oscillations.

4. State the problem involved in the injection of fuel into the cylinder of a modern heavy oil engine, explaining how it is carried out in (*a*) the pure Diesel type using blast air, and (*b*) the airless injection type: also indicate how far the complete solution of the problem is achieved.

Shew that the energy spent in diffusing the oil in (*a*) with blast air at a pressure of 1000 lb. per sq. in., temperature 60° C. and volume 5 per cent. of the main air is about four times that in (*b*) with 3000 lb. per sq. in. oil pressure. The compression pressure is 500 lb. per sq. in. Oil is found to require about fifteen times its weight of air for combustion, but in the exhaust of a Diesel engine about half the oxygen is found unused. [Specific gravity of oil 0·85.]

5. In an internal combustion engine, the volume of fresh charge drawn in during each suction stroke when reckoned at atmospheric temperature (15° C.) and pressure (14·7 lb. per sq. in.) is 0·9 of the swept volume (measured between the two piston positions at which the pressure in the cylinder is atmospheric). The temperature of the residual exhaust gases may be taken as 700° C. Find the temperature of the cylinder contents at atmospheric pressure on the compression stroke.

Take the ratio of swept cylinder volume (as defined above) to residual exhaust volume as 4 : 1.

What would be the effect of taking the temperature of the residual exhaust gases as 800° C.?

What factors influence the volumetric efficiency of an internal combustion engine?

6. A four-stage air compressor is designed to run at constant speed, taking in air at 15 lb. per sq. in. and delivering it at 1215 lb. per sq. in. Intercooling takes place down to atmospheric temperature at each interstage, and the work done in each compression stage is the same. The compression and expansion lines in the indicator diagrams for all stages may be taken as always of the form $pV^{1 \cdot 3}=\text{constant}$. The ratio ($c$) of clearance volume to stroke volume may also be taken as the same in each stage, but to effect volume control of the air delivered, this ratio may be altered to suit the load. Assuming that at full load c is negligible, shew that at the proportion $x:1$ of full load, the proper value of c is approximately

$$c=0 \cdot 75\,(1-x).$$

7. In a trial of a CO_2 refrigerator, in which the brine is circulated by a pump, the following readings were taken:

Duration of trial: 2 hours.

Total I.H.P.: 8·82.

Mean temperature of brine (evaporator inlet): −12·05° C.

„ „ „ „ („ outlet): −17·73° C.

Total gallons of brine circulated: 1000.

Mean inlet temperature of condenser water: 20·55° C.

„ outlet „ „ „ „ : 23·16° C.

Total gallons of condenser water used: 2520.

Radiation loss from plant to surroundings per hour (measured): 3830 C.H.U.

Thermal capacity of brine per gallon: 8·50 C.H.U.

Neglecting the work required to pump the brine, draw out a heat balance sheet, and find the coefficient of performance.

Give an outline diagram of the plant.

8. Superheated steam at a pressure of 180 lb. per sq. in. and temperature 250° C. is expanded through a converging nozzle to a pressure of 120 lb. per sq. in. The velocity efficiency of the nozzle is 95 per cent. Determine the state of the issuing steam, and the gain of entropy.

9. In the first pressure stage of a turbine of the Curtis type, there are two velocity stages. The steam velocity at the nozzle is 2250 ft. per sec., whilst the mean peripheral speed of the moving blades is 450 ft. per sec. In its passage through the fixed blades, the steam may be assumed to have its velocity unaltered, but in passing through the moving blades, the relative steam velocity may be assumed to be diminished by 20 per cent. in each case. The steam jet angle is 20°, the exit angles of the blades being: (i) first moving, 22°, (ii) fixed, 24°, (iii) second moving, 35°. Find the inlet angles, the velocity in magnitude and direction of the exhausting steam, and the work done per lb. of steam.

Calculate how the blade heights vary (in terms of the nozzle height), neglecting the variations of the quality of the steam due to friction.

10. The quantity of heat which can be transmitted per sq. ft. per hour from the hot gases to the water in a boiler may be assumed to vary as the square of the difference in temperature

between the gases and water. On this basis, shew that the efficiency of the heating surface of a boiler is

$$\frac{S}{S+\frac{\sigma W}{c\theta}},$$

in which S is the heating surface,

W the weight of gases in lb. per hr.,

σ the specific heat of the gases,

θ the difference in temperature between the gases at the furnace end of the tubes and the water in the boiler,

and c the heat transmitted per hr. per sq. ft. per degree difference of temperature.

PAPER XLV

1. A working fluid goes through a Carnot cycle of operations, the upper absolute temperature of the fluid being θ_1, the lower absolute temperature being θ_2. The amounts of heat per sec. taken in and rejected by the fluid are H_1 and H_2 respectively. On account of heat conduction effects, differences of absolute temperature exist between the source (T_1) and θ_1, and between θ_2 and the sink (T_2). If $T_1-\theta_1=\kappa H_1$, $\theta_2-T_2=\kappa H_2$, where κ is the same constant for both temperature differences, shew that the efficiency of the plant is

$$1-\frac{T_2}{T_1-2\kappa H_1}.$$

2. Two vessels A and B contain equal masses of gaseous mixture at temperatures of 1440° C. and 288° C. respectively. A reversible engine takes in heat from A and rejects heat to B. The specific heat at constant volume of the mixture is $(0{\cdot}17+4{\cdot}5\times10^{-5}\,T)$, where T is the absolute temperature. Shew that when the engine has done the greatest amount of work possible the temperature of A and B will be about 677° C. Find the work done per lb. of mixture in A or B.

3. The temperature θ degrees centigrade of a coal calorimeter during heating is given by $\theta=15e^{-\frac{t}{11}}(t+1)$, where t is the time in sec. Find the maximum temperature attained during the experiment.

During cooling the temperature fell uniformly from 61° C. to 59° C. in 10 sec. and the radiation may be assumed proportional to the temperature difference between the calorimeter and the room, which was at 15° C. Shew that the radiation correction up to the time of maximum temperature is 1·63° C.

4. In a plant in which the demand for steam is intermittent or subject to considerable variation, a closed vessel is placed on the top of the boiler and filled for the most part with water. The boiler, being of smaller capacity than would be required for the top load, is kept working at full power, and when all its steam is not required for the engines, the surplus steam is blown into the vessel of water. Subsequently when the steam required is beyond the capacity of the boiler, steam is obtained by gradually lowering the pressure upon the vessel of water. State the advantages of this method, and estimate whether economy is likely to result.

The boiler is working at a pressure of 150 lb. per sq. in. and the vessel contains 3000 gallons of water at 120° C. Steam is blown into the vessel direct from the boiler and the temperature of the water raised to the temperature of the steam. Find how many pounds of steam at 120° C. can be formed by gradually lowering the pressure in the vessel until the water is at 120° C.

Compare the work that may be expected from this steam in an engine expanding to a pressure of 2 lb., with the work to be obtained from the steam which was fed into the vessel, if instead the steam had been taken direct from the boiler to the engine.

5. The clearance space of a steam engine cylinder is 2 cu. ft. At the end of compression the pressure is 15 lb. per sq. in. and the dryness fraction of the mixture in the clearance space is 0·95. The steam from the boiler is at a pressure of 100 lb. per sq. in. and its dryness fraction is 0·9. Assuming the walls to be non-conductors of heat and that no loss of pressure takes place between the boiler and the cylinder, estimate (1) the quantity of steam admitted into the clearance space up to the moment at which the piston just commences to move forward, and the dryness fraction of the mixture in the clearance space at this

stage, (2) the dryness fraction of the mixture after 1 lb. of boiler steam has entered the cylinder.

6. The compression ratio for a petrol engine is 5 : 1, and the theoretical temperature at the end of the explosion period is estimated to be 3050° C. Calculate the theoretical temperature at the end of the expansion stroke, assuming the mean value of γ during this stroke to be 1·225.

The molecular heat at constant volume, at temperature t° C., of the products of combustion is $5{\cdot}35 + {\cdot}001416t$. The universal gas constant may be taken as 1·985. Is the mean value of 1·225 taken for γ consistent with these facts?

7. In a trial of a boiler and economiser, the following observations were made:

Feed water per hour	3200 lb.
Coal fired per hour	450 lb.
Temperature of feed to boiler	57° C.
,, ,, ,, economiser ...	25° C.
,, ,, flue gases leaving boiler ...	310° C.
,, ,, ,, ,, economiser	146° C.
,, ,, boiler room	16° C.
Steam pressure	160 lb. per sq. in.
,, temperature	280° C.
Ash per hour	53 lb.
Calorific value of ash	1500 TH.U.
,, ,, coal	7500 TH.U.
Mean value of K_p for flue gases	0·24

There is no CO in the flue gases and the analysis of the coal and flue gas shews that 21·25 lb. of dry flue gas and 0·46 lb. of steam are produced per lb. of coal burnt.

Draw out balance sheets for the boiler and the economiser, estimated on 1 lb. of coal.

8. In two-stage air compressors, where the initial pressure is p_0 and the final pressure is rp_0 and the air passes through an intercooler between the cylinders so that its temperature is brought back to its initial value, it is customary to make the intermediate pressure equal to $r^{\frac{1}{2}}p_0$. Justify this practice in a

compressor with no clearance and with compression following the same law in each stage, and shew that there is the same rise of temperature in each cylinder.

Find the ratio of the diameters of the cylinders of a two-stage tandem air compressor in which the initial and final pressures are 15 lb. per sq. in. and 495 lb. per sq. in. respectively, and the intermediate pressure is given by the rule stated in the first part of the question.

The ratio of clearance volume to stroke volume is $\frac{5}{100}$ for the L.P. cylinder and $\frac{8}{100}$ for the H.P. cylinder, and the compressions and expansions follow the law $pV^{1\cdot 3}=\text{constant}$.

The atmospheric pressure is 15 lb. per sq. in.

9. Steam is discharged through a nozzle from a boiler in which a pressure of 120 lb. per sq. in. is maintained. The initial dryness of the steam is 0·9. The rate of discharge is 20 lb. per min. and the pressure is 60 lb. per sq. in. at a point in the nozzle where the diameter is 0·5 in. Find the state of the steam and its velocity at this section and determine what fraction of the available energy has been lost.

10. A certain impulse turbine is divided into two stages, each with two sets of moving blades. Steam at 180 lb. per sq. in., superheated 60° C., is supplied to the first stage and the condenser vacuum is maintained at 28 in. of mercury. The pressure at the end of the first stage is to be that corresponding to an equal division of the total adiabatic heat drop between the stages. The nozzles at the beginning of each stage are inclined at 20° to the plane of rotation of the blades. The moving blades have a mean speed of 400 ft. per sec.

Taking the loss of kinetic energy by friction in the nozzles as 15 per cent. of the available heat drop, and a 20 per cent. loss of velocity for each fixed and moving blade, determine, for the first stage only, the angles of entry and exit (which are to be made the same) for the two sets of moving and one set of fixed blades.

What is the 'carry over' velocity of the steam between the first and second stages?

PAPER XLVI

1. A cartridge containing 4 lb. of air at 1000 lb. per sq. in. by gauge and 15° C. is placed in the chamber of a gun behind a light frictionless piston fitting the bore of the gun. The cartridge is perforated and the piston just reaches the muzzle of the gun. Calculate the mean temperature of the air and the volume of the gun on the assumption that the air absorbs no heat from the walls of the gun. Atmospheric pressure = 14·7 lb. per sq. in.

2. In a reversible heat engine, between the source and the engine and between the engine and receiver, there are conduction and radiation effects. The working substance receives and rejects its heat at temperatures of 220° C. and 60° C. respectively and 10 per cent. of the heat from the source and 10 per cent. of the heat leaving the engine are lost by radiation.

Find the actual thermal efficiency of the plant and draw up a heat balance sheet shewing the distribution of 1000 TH.U. leaving the source.

3. For engines which are to be worked intermittently heat is accumulated in large cylinders filled with water. When the engines are at rest the boilers are employed in raising the temperature of the water in the cylinders, and the stored heat produces steam when the pressure on the water is diminished. The highest pressure to which the cylinders are exposed is 20 lb. per sq. in., and this pressure is reduced to 10 lb. per sq. in. when steam is required. Determine how many cubic feet of water will be required in the storage cylinders to drive turbines of 10,000 horse power for 40 min. with a consumption of 12 lb. of steam per horse-power hour.

4. The figure shews a portion of the indicator diagram for a four-cycle internal combustion engine. *ABC* represents the suction stroke and *CDE* a portion of the compression stroke. Scale of pressures, 1 in. = 5 lb. per sq. in. Compression ratio, 4·5. Atmospheric pressure, 14·7 lb. per sq. in. Temperature of residual gases, 550° C. The specific heat of the residual gases is assumed to be the same as that of the incoming charge. $\gamma = 1{\cdot}4$.

Taking the temperature of the incoming charge as 25° C. and neglecting the heat abstracted from the cylinder walls, find the temperature of the cylinder contents at the point D on the compression stroke.

The method employed in obtaining the result should be carefully explained.

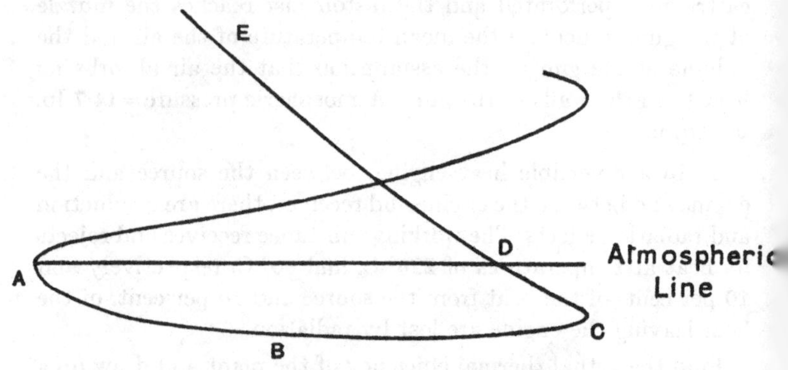

5. Shew that the energy spent in diffusing the oil-feed to an engine, (*a*) with blast air at a pressure of 1000 lb. per sq. in., temperature 60° C. and volume 5 per cent. of the main air, is about four times that in (*b*) with 3000 lb. per sq. in. oil pressure. The compression pressure is 500 lb. per sq. in. Oil is found to require about fifteen times its weight of air for combustion, but in the exhaust of a Diesel engine about half the oxygen is found unused. [Specific gravity of oil, 0·85.]

6. During a trial of one of the large air refrigerating machines used at the London Docks for maintaining cold storage the figures given below were obtained. The machines work by taking in air from the cold chamber and returning it again after compression, cooling, and expansion. The air in these machines, after the usual compression and cooling, is further cooled to just above 0° C. in order to deposit as much of the suspended moisture as possible before entering the expansion cylinder. This is done by passing it through an 'interchanger' in which it gives up heat to the cold air which is on its way from the cold chamber to the compressor.

Power absorbed in driving the pumps 303 H.P.
Quantity of air circulating per hour ... 170,000 cu. ft. at N.T.P.

Temperature in the cold room		$-8°$ C.
,,	of air entering compressor	$+12°$ C.
,,	,, after compression	144° C.
,,	,, after cooler ...	18° C.
,,	,, after interchanger	3° C.
,,	,, after expansion ...	$-58°$ C.

Calculate the actual coefficient of performance, and the ideal coefficient for an air machine working on the reversed Joule cycle with the same maximum and minimum temperatures as those in the test but without the extra cooling before expansion given by the interchanger.

What was the rate of leakage of heat into the interchanger from outside?

The specific heat of the air at constant pressure, allowing for the presence of moisture, may be taken as 0·25.

7. Two refrigerating machines, one using sulphur dioxide and the other ammonia, work between the same limits of temperature, viz. 10° C. to $-35°$ C. In each case the vapour is dry and saturated at the end of compression, and expansion takes place through a throttling valve. Determine in each case the refrigerating effect per lb. of substance used and the coefficient of performance.

Also shew that if the machines are to have the same refrigerating effect per stroke, the volumes of their compressing cylinders must be approximately in the ratio of 11 to 4.

8. In a pressure stage of an impulse turbine, steam expands adiabatically in the nozzles from 50 lb. per sq. in. and 32° C. superheat to 12 lb. per sq. in. The expansion after reaching normal saturation point continues according to the law $pv^{1\cdot3}$ = constant, owing to supersaturation. Find the loss due to this action as compared with the full adiabatic heat drop, and determine the velocity of exit.

Describe and account for the change in form of the nozzles according as the ratio of the lower pressure to the upper pressure is below (as in this case) or above the critical ratio

$$\left(\frac{2}{\gamma+1}\right)^{\gamma/(\gamma-1)}=0\cdot54.$$

9. Shew that if per lb. of steam used in a steam turbine running steadily W be the shaft work and H the 'external' loss (including radiation, shaft friction, leakage and kinetic energy of exhaust steam), then the heat balance is given by:

$$\text{Heat drop} = I_1 - I_2 = W + H,$$

where I_1 and I_2 are the total heats per lb. of the supply and exhaust steam respectively. All quantities are measured in the same units.

A steam turbo-alternator develops 10,000 KW. at 1500 R.P.M., the generator efficiency being 96 per cent. The steam supply is at 180 lb. per sq. in. (gauge), temperature 288° C. The vacuum is $28\frac{1}{2}$ in., the barometer being 30 in., and the steam consumption per hour is 122,000 lb. Taking the outside losses at 5 per cent. of the frictionless adiabatic heat drop, find per lb. the actual heat drop and the volume of the exhaust steam. Thermal equilibrium is to be assumed.

If the last row of blades has a mean diameter of 72 in., height of blades 14 in., exit angle of 35°, find the absolute exit velocity of the steam, and the kinetic energy per lb. corresponding, estimated as a percentage of the actual heat drop.

10. A long straight cylindrical pipe, of internal radius r_1 and external radius r_2, has its inner surface maintained at temperature θ_1 and its outer surface at temperature θ_2. If θ is the temperature at an intermediate radius r, shew that

$$\theta = \frac{\theta_1 \log_e \dfrac{r_2}{r} + \theta_2 \log_e \dfrac{r}{r_1}}{\log_e \dfrac{r_2}{r_1}}.$$

The external diameter of a long straight pipe is 0·75 in. and the thickness 0·05 in. The temperatures of the outer and inner surfaces are maintained at 39° C. and 16° C. respectively. The specific conductivity of the material of the pipe is 0·22 in C.G.S. units. Find the amount of heat which passes through the walls of the pipe per ft. length per min.

PAPER XLVII

1. A non-conducting vessel is partitioned by a diaphragm and contains one lb. of a gas in each compartment. The pressure throughout the vessel is 15 lb. per sq. in. but the temperatures on the two sides are 1000 and 2000° C. absolute respectively. Determine the final temperature and pressure inside the vessel if the partition breaks down:

(*a*) if the specific heats are constant;

(*b*) if the specific heats vary with the absolute temperature T, such that

$$K_v = 0{\cdot}1680 + 4{\cdot}5 \times 10^{-5} T, \text{ in thermal units per lb.}$$

2. Two homogeneous bodies of mass M and specific heat S and differing in temperature by a small quantity t are placed in contact with one another, until they have one common temperature T. Shew that the increase in entropy of the bodies is very approximately $\dfrac{MSt^2}{4T^2}$.

3. Explain Mollier's ϕ-I diagram and shew how the lines of the diagram are set out.

By means of the diagram or otherwise solve the following cases:

(i) In a compound engine of two cylinders the highest pressure is 150 lb. per sq. in. and the lowest 2 lb. per sq. in. The steam is dry at entry and expands to 30 lb. per sq. in. in the high pressure cylinder and on leaving the cylinder is throttled before entering the low pressure cylinder. If the engine works upon the Rankine cycle and the work from both cylinders is the same, determine the range of pressure in the low pressure cylinder and the efficiency of the engine. Determine the increase of entropy of the steam during throttling.

(ii) Dry steam entering a turbine at 150 lb. per sq. in. leaves the turbine at a pressure of 1 lb. per sq. in. and dryness 0·85. Determine the increase in entropy. If during the process the increase in entropy has been in proportion to the fall in I, and the fall in I is the same for each of the three stages of the turbine, determine the pressure and dryness at the end of the first and second stages.

4. Water is flowing steadily through a straight cylindrical tube of uniform cross-section, which is surrounded by steam at constant temperature. Assuming that the heat flow per second per unit length of tube is always proportional to the temperature difference θ between the steam and water, shew that the mean value of θ (averaged over the whole length of tube) is given by

$$\theta_m = \frac{\theta_1 - \theta_2}{\log \frac{\theta_1}{\theta_2}}.$$

In this expression, θ_1 and θ_2 are the temperature differences between the steam and water at entry and exit respectively of the water to the tube.

If the steam temperature is 30° C., the inlet and outlet temperatures of the water 15° C. and 25° C. respectively, find the mean temperature difference θ_m, and compare it with the arithmetic mean of the temperature differences (θ_1 and θ_2) at inlet and exit of the water.

5. The Atkinson cycle, in which heat is supplied at constant volume and rejected at constant pressure, has been recommended as a standard of reference for internal combustion engines. Comment on this by considering the relation of its diagram to that of (*a*) the Otto or constant volume cycle, (*b*) the Diesel cycle.

For the Otto constant volume cycle shew that the mean effective pressure is given by

$$pr \frac{(\alpha - 1)(r^{\gamma-1} - 1)}{(\gamma - 1)(r - 1)},$$

where p is pressure at beginning of compression, r is compression ratio, α is ratio of maximum pressure to compression pressure.

Hence compare the output from the same engine using (*a*) gas of calorific value 280 TH.U. per cu. ft., requiring 7 times its volume of air for combustion, and (*b*) gas of 56 TH.U. calorific value, requiring its own volume of air.

6. In a Diesel engine, oil of calorific value 10,000 TH.U. is burnt at constant pressure in the proportion of 1 lb. oil to 28 lb.

air, the resulting products having a specific heat varying with the absolute temperature T according to the relation

$$K_p = 0{\cdot}24 + 4{\cdot}5 \times 10^{-5}\, T.$$

The compression ratio is 13·5, and the temperature after compression may be taken as 800° C. Find the percentage of stroke at which cut-off occurs.

Indicate briefly an approximate method of allowing for variable specific heat in applying calculations during the subsequent expansion.

7. The following particulars relate to a boiler trial:

Feed water per hour = 9000 lb.
Wet coal per hour (as fired) = 1200 lb.
Moisture in coal (as fired) = 8 per cent.
Lower calorific value per lb. of dry coal = 8000 C.H.U.
Analysis by weight of dry coal:
$C = 84$ per cent., $H_2 = 5$ per cent., $O_2 = 7$ per cent.,
$N_2 = 1$ per cent., Ash = 3 per cent.
Volumetric analysis of the dry flue gases gave

$$CO_2 : O_2 = 3 : 4.$$

Steam pressure = 200 lb. per sq. in. abs.
Temperature of superheated steam = 250° C.
Temperature of feed water = 50° C.
Temperature of boiler house = 15° C.
Temperature of flue gases leaving superheater = 265° C.
Specific heat at constant pressure of the flue gases = 0·24.

Find the combined thermal efficiency of the boiler and superheater and the weight of air supplied per lb. of wet fuel in excess of that required for complete combustion.

What percentage of the heat supply is carried away by the flue gases?

Estimate the higher calorific value of the dry coal.

[Ratio of nitrogen to oxygen in air by weight, 77 : 23.]

8. The figures below refer to a one hour's trial of an ammonia refrigerating machine working with single stage compression and with an expansion valve between the condenser and evaporator:

Average I.H.P. of compressor	15·2 H.P.
Evaporator pressure	41·5 lb. per sq. in.
Condenser pressure	130 lb. per sq. in.

Brine, quantity per hour	5055 gallons.
Brine temperature, entering evaporator ...	−2° C.
Brine temperature, leaving evaporator ...	−5° C.
Cooling water, quantity per hour	1500 gallons.
Cooling water temperature, entering condenser	9·5° C.
Cooling water temperature, leaving condenser	19·7° C.

The specific heat of the brine was 8·45 TH.U. per gallon per ° C. rise of temperature.

Draw up a heat balance sheet and determine the coefficient of performance. Compare the actual coefficient of performance with that which would be given by an ideal refrigerating machine, working with adiabatic compression and an expansion valve, the limits of pressure being those of the actual machine, the vapour being just dry at the end of compression and the liquid cooled in the condenser to 10° C.

[1 gallon of water weighs 10 lb.]

9. In an outward radial flow turbine consisting of alternate fixed and moving rings of blades, the latter rotating with angular velocity ω, the blading is of the reaction type, of exit angle θ for both fixed and moving rings and so arranged that the ratio

$$\frac{\text{blade velocity at entrance}}{\text{steam velocity (abs.) at entrance}} = \frac{\text{blade velocity at exit}}{\text{steam velocity (rel.) at exit}} = \rho,$$

i.e. the velocity triangles at entrance and exit of the moving blades are similar.

Shew that the work done on a moving ring of internal and external radii r_1 and r_2 respectively is:

$$\frac{\omega^2}{g}\left\{\frac{\cos\theta}{\rho}(r_1^2+r_2^2)-r_2^2\right\} \text{ per lb. of steam.}$$

If, as in the Ljungström turbine, both sets of rings rotate in opposite directions, explain what particular difficulty in turbine design is thereby overcome, mentioning two methods of dealing with the same problem in axial flow turbines.

10. Explain the meaning of *stage efficiency* in a multi-stage turbine.

In consecutive stages in a pressure-compounded turbine, the initial pressures are 150, 90, 55, and 30 lb. per sq. in., and the stage efficiency is 0·7 for each stage; if the initial temperature of the steam is 300° C., find its condition at the beginning of each stage and the entropy per lb. on entering the last stage.

Shew how the adiabatic heat drop for the whole turbine is related to the work done and the stage efficiency by a 'reheat' factor, and hence that over-all efficiency, defined as

$$\frac{\text{work done on rotor}}{\text{adiabatic heat drop}},$$

is greater than the stage efficiency.

PAPER XLVIII

1. A vessel of volume V contains initially air at a pressure p_1, the walls being insulated to prevent the passage of heat. Air is then admitted from a supply at pressure p by a pipe provided with a stop valve until the pressure in the vessel is equal to p.

Shew that the weight of air which enters the vessel from the supply is

$$\rho . V \frac{1}{\gamma} \frac{p - p_1}{p},$$

in which ρ is the density of the air in the supply.

If the pressure and temperature are initially 15 lb. per sq. in. and 15° C. respectively in the vessel and 100 lb. per sq. in. and 20° C. in the supply, what is the final temperature of the air in the vessel? $\gamma = 1{\cdot}41$ for air.

2. During the expansion of steam in the cylinder of an engine an interchange of heat Q between the cylinder walls and the steam takes place which may be represented by:

$$\frac{dQ}{dT} = a + bT,$$

where T is the absolute temperature of the steam and a and b are constants.

If dry saturated steam at 200° C. expands to 40° C. and if 0·3 TH.U. per 1° C. fall are given out by the steam at the upper temperature and 0·6 TH.U. per 1° C. fall are received by it at the lower temperature, determine the final dryness of the steam and the amounts of heat given out and received by the steam during the expansion.

3. An engine working on the Rankine cycle is supplied with steam at 130 lb. per sq. in. and dryness 0·9, the pressure in the condenser being 2 lb. per sq. in. Determine its thermal efficiency.

In another engine, using steam at the same pressure and dryness and with the same condenser pressure, compound expansion in two cylinders is employed. The steam leaves the high pressure cylinder at 25 lb. per sq. in., and as it passes to the low pressure cylinder at the same pressure gives up a portion of its heat to the feed water. If the amount of heat so given up is sufficient to heat the feed water from the temperature of the condenser to that of the steam as it leaves the high pressure cylinder, find the thermal efficiency of the engine and the dryness of the steam as it enters the low pressure cylinder. The expansions in both cylinders are adiabatic and complete.

Shew that by an extension of this method of feed-water heating it is theoretically possible to make the efficiency of the engine equal to that of the Carnot engine.

4. Water at temperature θ_1 passes into a tube of length l and diameter d, and leaves the tube with a temperature θ_2; the temperature of the tube itself being θ_0. The heat given by the tube to the water is $\alpha\rho w\,(\theta_0-\theta)$, where α is a constant. ρ=the density of the water, w=its velocity, and θ=its temperature. Shew that the rise of temperature is given by

$$\theta_2-\theta_1=(\theta_0-\theta_1)\,(1-e^{-\mu}),$$

where $\mu=\frac{4\alpha l}{\sigma d}$, σ being the specific heat of the water.

5. A receiver vessel of volume V containing air initially at atmospheric pressure (p_a) and temperature (T_a) is pumped up to a pressure p by a small compressor working adiabatically and drawing steadily from the atmosphere a volume v per unit of time. If the temperature of the air in the receiver is maintained constant and equal to the atmospheric temperature T_a, shew that the time required to attain a pressure p is

$$t=\frac{V}{v}\left(\frac{p}{p_a}-1\right),$$

and the work done by the compressor is

$$W=\frac{\gamma}{\gamma-1}V\left[\frac{\gamma}{2\gamma-1}p_a\left\{\left(\frac{p}{p_a}\right)^{\frac{2\gamma-1}{\gamma}}-1\right\}-(p-p_a)\right],$$

γ being the ratio of the specific heats.

6. The figure shews a light-spring indicator diagram taken from the crank-case of a two-stroke cycle oil engine of the type

in which the underside of the working piston is used to compress the scavenge air. The diameter of the cylinder is 9 in., the stroke 17 in., and the compression ratio 7.

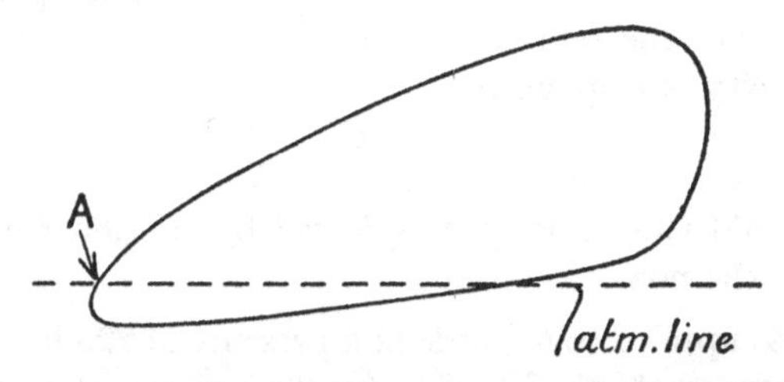

In the process of scavenging, which may be assumed to take place at substantially atmospheric pressure with the piston stationary at the end of its stroke, the scavenge air mixes with the expanded gases (themselves at 500° C. and consisting as to 30 per cent. of unburnt air) driving out a portion, but losing to exhaust a proportion e^{-x} of itself, where x is the amount injected measured in cylinder volumes.

If the temperature of the air at *A* on the diagram, and also during scavenge is estimated at 35° C., find (*a*) the amount of scavenge air supplied and retained per revolution, (*b*) the temperature of the final charge and the amount of oil which theoretically could be burnt. Oil/air $=\frac{1}{15}$ by weight for complete combustion. Specific heats are to be taken as the same throughout.

Why is this theoretical value not realised in practice?

7. Explain the reason for keeping the clearance volume as small as possible in reciprocating air compressors.

The following results were obtained in a test of a twin-cylinder single-stage, single-acting air compressor, running at 391 R.P.M.

Delivery gauge pressure, lb. per sq. in.	25	50	75	100
Volume of free air compressed per min. —cu. ft. ...	29·2	28·8	28·1	27·2

Cylinders: 4 in. stroke, 5 in. bore. Clearance: 3 per cent. of swept volume.

Plot on a delivery gauge pressure base the curves of apparent and true volumetric efficiency.

The suction pressure may be taken as atmospheric (14·7 lb. per sq. in.) and the expansion to follow the law $pv^{1\cdot 3}=\text{const.}$

8. Shew that, for the adiabatic flow of steam along a properly shaped nozzle, the relationship between I, volume v, pressure p and velocity w is given by

$$-dI=-\frac{v\,dp}{J}=\frac{w\,dw}{Jg}.$$

How would the velocity be affected by friction between the steam and the nozzle?

Steam is supplied to a nozzle at a pressure of 225 lb. per sq. in. and at a temperature of 400° C. (or 200° of superheat), and expands down to a pressure of 150 lb. per sq. in. Owing to friction in the nozzle the kinetic energy at discharge is 15 per cent. less than with adiabatic and frictionless flow. Determine the state and velocity of the steam leaving the nozzle.

If the characteristic equation for superheated steam is

$$v-0\cdot016=1\cdot07\,\frac{T}{p}-0\cdot42\left(\frac{373}{T}\right)^{\frac{10}{3}},$$

where v is volume in cu. ft. per lb., p is pressure in lb. per sq. in., T is absolute temperature, determine the exit area necessary for a discharge of 10 lb. per sec.

9. The diagram given (fig. 1) shews the first pressure stage and two velocity stages of an impulse turbine.

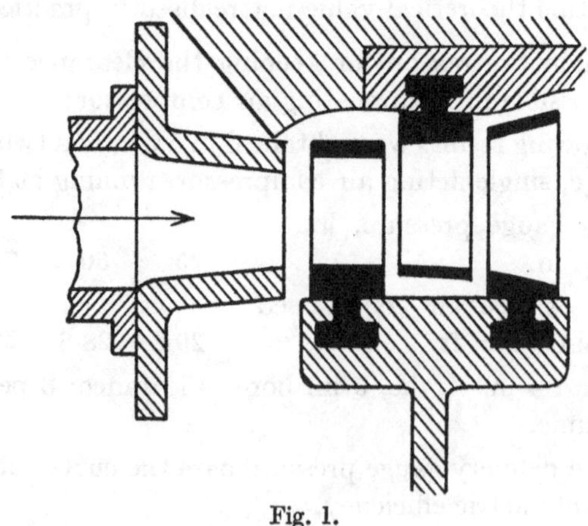

Fig. 1.

The nozzle is set at an angle of 20° and the steam issues at a speed of 2500 ft. per sec. If the velocity coefficient in the blade passages is 0·9, and the exit angle of the blade passages 23°, determine for a blade speed of 550 ft. per sec.

(i) The work done per lb. of steam.

(ii) The efficiency of the stage, if the steam has fallen in the nozzle from a pressure of 200 lb. per sq. in. and temperature 300° C. to a pressure of 35 lb. per sq. in. and a temperature of 130° C.

Graphical methods may be used.

10. Of each 100 thermal units supplied to an oil engine, 30 are found in B.H.P., 25 in cylinder cooling water, 30 in exhaust gas calorimeter water and the remaining 15 in chimney gases, radiation, friction, etc.

The jackets and exhaust gas calorimeter of the engine are connected up so that they are made part of the heating surface of a boiler as shewn in fig. 2. The boiler works at 130 lb. per sq. in. and may be taken to be of efficiency 70 per cent. The steam

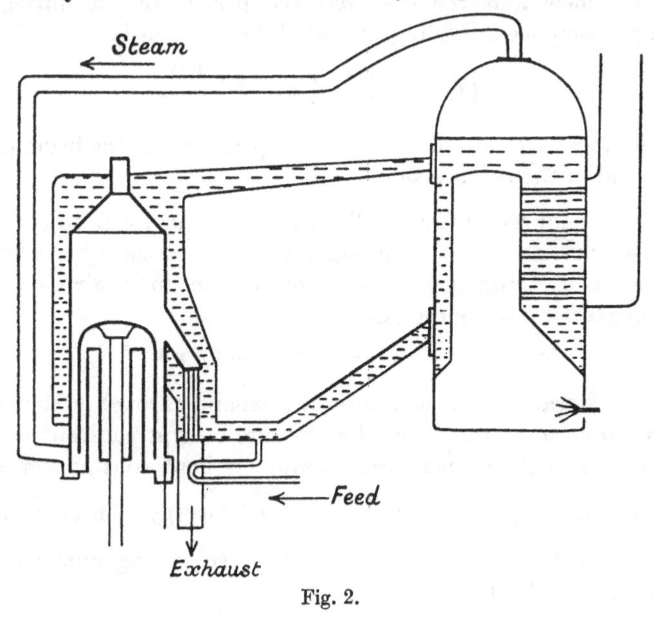

Fig. 2.

formed is admitted to the underside of the piston of the oil engine, adding to the power of the engine.

If with a vacuum pressure of 2 lb. per sq. in. 18 lb. of steam are required to contribute 1 B.H.P. hour, find the brake-efficiency of the combined oil and steam engine:

(i) When 1 lb. of oil is burnt in the boiler for every 10 lb. of oil used in the engine.

(ii) When no oil is burnt in the boiler.

The calorific value of the oil is 10,000 TH.U. per lb.

PAPER XLIX

1. For a perfect gas, whose specific heats are given by the equations $K_v = a + sT$ and $K_p = b + sT$, in which a, b and s are constants, shew that during an adiabatic compression $pV^\gamma e^{\lambda T}$ is constant, where

$$\gamma = \frac{b}{a} \text{ and } \lambda = \frac{s}{a}.$$

Also shew that the heat received per lb. of gas during a compression according to the law $pV^n = \text{constant}$ is

$$a\,(T_2 - T_1)\left[\frac{n-\gamma}{n-1} + \lambda\,\frac{T_2 + T_1}{2}\right],$$

where T_1 and T_2 are the absolute temperatures at the beginning and end of the compression.

Air is compressed from 15 lb. per sq. in. and 15° C. to 600 lb. per sq. in. and 450° C. according to a law of the type $pV^n =$ constant. Determine the amount of heat per lb. of air received or rejected in the compression.

[For air $a = 0{\cdot}17$, $b = 0{\cdot}24$, $s = 4{\cdot}5 \times 10^{-5}$.]

2. A stream of gas is allowed to expand through a throttle valve from a region where the pressure is kept constant at p_1 to another region where the pressure is kept constant at p_2. Shew that the process entails a gain of entropy of $R \log \frac{p_1}{p_2}$ per unit mass, if the gas is perfect and the conveying pipe a non-conductor of heat.

3. Steam of pressure 225 lb. per sq. in. and temperature 400° C. expands in an engine to a pressure of 40 lb. per sq. in., and is then reheated at this pressure to the temperature 400° C. It is then expanded further in the engine to the condenser pressure of 1 lb. per sq. in. Shew that the introduction of the reheating stage increases the efficiency of the engine by approximately 5 per cent.

The expansion of the steam in the engine may be assumed adiabatic and of constant entropy.

4. The exhaust steam from a steam-engine cylinder is used to vaporise liquid SO_2, and the SO_2 is expanded in a second cylinder. Assuming that the stuff in both cylinders is just dry at the point of cut-off, that clearances etc. are neglected, and that expansion, in each case, is complete and adiabatic, estimate

(1) the efficiency of the steam engine alone,

(2) the number of lb. of SO_2 used in the vapour cylinder per lb. of steam used in the steam cylinder,

(3) the dryness fractions of the steam and SO_2 at the end of expansion,

(4) the combined efficiency; having given

initial temperature of steam ... ... =165° C.
exhaust temperature of steam ... ... =35° C.
initial temperature of SO_2 ... ... =30° C.
exhaust temperature of SO_2 ... ... =15° C.
average specific heat of liquid SO_2 between 30° and 15° = 0·4.

The liquid SO_2 may be supposed to enter the steam-engine condenser at 30°, and the condensed water from the steam cylinder to leave it at 35°.

5. Shew that the constant R per lb. for a gas in the law $pV=RT$ is inversely proportional to the molecular weight, and hence that in a mixture of two gases:

$$\frac{p_1}{p}=\frac{m_1M_2}{m_2M_1+m_1M_2},$$

where M_1, M_2 are the molecular weights of each constituent,

m_1, m_2 are the masses of each constituent,

p_1 is the partial pressure of one constituent,

P is the total pressure.

In distilling an oil, steam is blown in, and the outlet from the still runs 10 per cent. water and 90 per cent. oil by weight, the process being carried out at atmospheric pressure and normal barometer. If the boiling point of the oil is 270° C. at normal atmospheric pressure and is lowered uniformly 3·8° C. for each lb. per sq. in. fall of pressure, shew that distillation occurs at a temperature of 238° C., assuming the vapours behave as gases.

[Molecular weight of the oil = 212, of steam = 18.]

6. A perfect gas ($pV = RT$ per lb.) has specific heats at temperature T abs. given by $\kappa_v = \alpha + \lambda T$, $\kappa_p = \beta + \lambda T$, where α, β and λ are constants, and $R = \beta - \alpha$. Shew that along an adiabatic

$$p^{\alpha} V^{\beta} e^{\lambda T} = \text{constant}.$$

A charge of this gas is compressed adiabatically in an engine from (p_1, T_1) to (p_2, T_2). If $\lambda (T_2 - T_1)$ is small, shew that approximately

$$T_2 = \frac{T_1 (\beta + \lambda T_1)}{\beta x + \lambda T_1},$$

where

$$x = \left(\frac{p_2}{p_1}\right)^{\frac{\alpha - \beta}{\beta}}.$$

7. In a trial of a boiler and economiser, the following figures were obtained:

Feed water per lb. of coal	6·90 lb.
Temperature of feed to boiler	144° C.
,, ,, ,, economiser	40° C.
,, ,, flue gases leaving boiler ...	374° C.
,, ,, ,, ,, ,, economiser	192° C.
,, ,, boiler room	14° C.
Steam pressure (lb. per sq. in. abs.) ...	150
,, temperature	250° C.
Ashes from grate per lb. of coal	0·013 lb.

Calorific value of ash per lb.	... ...	2000 C.H.U.
,, ,, coal ,, (higher)	...	6320 C.H.U.
Mean value of K_p for flue gases	... ...	0·25

There is no CO in the flue gases, and the analyses of coal and flue gas shew that 17·67 lb. of dry flue gas and 0·49 lb. of H_2O are produced per lb. of coal burnt.

Draw out a heat balance sheet (estimated on 1 lb. of coal) for (i) the boiler alone, (ii) the economiser alone. Neglect any leakage of air through the brickwork.

8. In the measurement of the air supply to an engine, through an orifice of the type shewn in fig. 1, the pressure difference on the two sides, as measured by a manometer, is found to vary with time, as shewn in fig. 2.

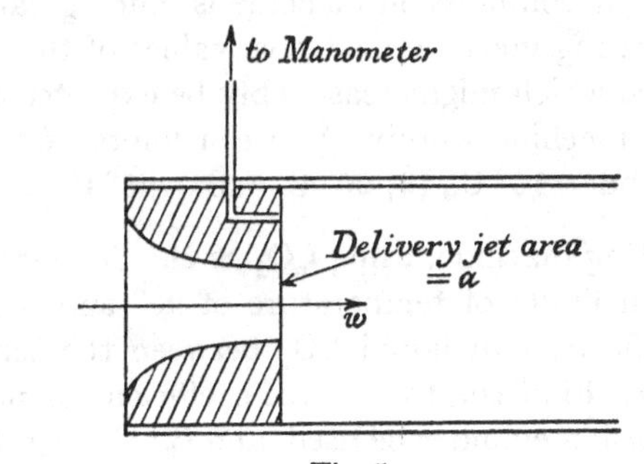

Fig. 1.

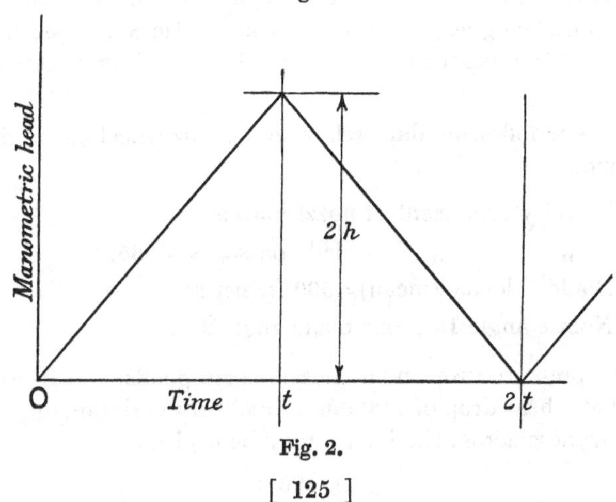

Fig. 2.

If the pressure variation is sufficiently slow for the velocities to conform to the usual expressions, shew that an error of just over 6 per cent. in the estimated volume passing will be introduced by taking the mean pressure difference instead of the true one.

If a is 2 sq. in. and h is 1 in. of water, determine the volume passing per min. Barometer 30 in. of mercury. Temperature 15° C. Specific gravity of mercury 13·6.

9. Explain precisely what practical and thermodynamical considerations determine the choice of the liquefiable vapour used in refrigerating machines, and compare the relative merits of SO_2, NH_3 and CO_2. Explain why the coefficient of performance of a vapour compression machine is much greater than of an air refrigerating machine, and give values of the coefficients of performance which might reasonably be expected in practice from an NH_3 machine working between limits of temperature of (1) 20° C. and −10° C., (2) 20° C. and −25° C.

A refrigerating machine, using CO_2 as the liquefiable vapour, works between limits of temperature of 20° and −20° C. The average specific heat of liquid CO_2 between the temperatures named is 0·94. Find the theoretical coefficient of performance (1) if an expansion cylinder be used, and (2) if a regulating cock be substituted for an expansion cylinder—the volume of the liquid CO_2 being neglected, and the adiabatic compression being such that the vapour is just dry at the end of compression.

10. The following data relate to a 1-row wheel of an impulse turbine:

Velocity coefficient of nozzles = 0·94.

" " blade passages = 0·85.

Blade velocity (mean) = 500 ft. per sec.

Nozzle angle 19°, exit blade angle 25°.

The pressure drop in the nozzle corresponds to a frictionless adiabatic heat drop of 14·0 C.H. units. Any variation of pressure and dryness across the blades may be neglected.

Find

(1) the theoretical (a) and actual (b) steam speeds from the nozzle;

(2) the ratio of blade speed to (1a), i.e. the velocity ratio;

(3) the work done, per lb. of steam, in the stage;

(4) the ratio of (3) to 14·0 C.H. units, i.e. the stage efficiency;

(5) the ratio of blade height at exit to nozzle height.

Graphical methods may be employed.

PAPER L

1. Establish the propositions (a) and (b).

(a) A mixture of perfect gases contains a mass m_1 of gas of gas constant R_1, a mass m_2 of gas of gas constant R_2, and so on. Assuming the law of partial pressures, shew that the gas constant of the mixture is

$$(m_1R_1+m_2R_2+\ldots)/(m_1+m_2+\ldots).$$

(b) Shew that the entropy ϕ of unit mass of perfect gas of gas constant R and specific heat (at constant volume) κ_v is

$$\phi=\int\kappa_v\frac{dT}{T}+R\log V+\text{const.}$$

In the above propositions, R refers to $pV=RT$ for unit mass of gas.

A rigid closed vessel of capacity 1 cu. ft. contains, separated from each other by a thin partition, 0·21 cu. ft. of oxygen, and 0·79 cu. ft. of nitrogen, each gas being at 14·7 lb. per sq. in. and temperature 15° C. The partition is removed without escape of gas, and inter-diffusion of the gases takes place. Shew that if there is no heat supplied to or withdrawn from the gases, when diffusion is complete the pressure and temperature of the mixture will be the same as those of the gases before mixing. Take $\gamma=1{\cdot}40$ for both gases.

Find the gain of entropy due to the diffusion process.

2. Two vessels A and B contain equal masses of gaseous mixture at temperatures of 1440° C. abs. and 228° C. abs. respectively. A reversible engine takes in heat from A and rejects heat to B. The specific heat at constant volume of the mixture is $(0{\cdot}17+4{\cdot}5 . 10^{-5}T)$, where T is the absolute temperature. Shew that when the engine has done the greatest amount of work possible the temperature of A and B will be about 677° C. abs. Find the work done per lb. of mixture in A or B.

3. A heat engine receives its heat as the temperature rises from T_1 to T_2 according to the law $\frac{dQ}{dT}=a$; and rejects its heat as the temperature falls from T_3 to T_4 according to the law $\frac{dQ}{dT}=b$; where a and b are constants and $\frac{T_2}{T_1}=\frac{T_3}{T_4}$. Shew that the efficiency is $1-\frac{b}{a}.\frac{T_4}{T_1}$, and that if the cycle consists of reversible processes only, a must be equal to b.

4. Steam of given pressure and dryness expands adiabatically and under full resistance. Shew how its dryness at any subsequent pressure in its expansion may be determined.

In engines which are to be worked for a short time only, heat is accumulated in large cylinders filled with water. When the engines are at rest the boilers blow steam through a throttle valve into the water in the cylinder, and the stored heat produces steam when the pressure on the water is diminished.

The highest pressure to which the cylinders are exposed is 20 lb. per sq. in. and this pressure is gradually reduced to 5 lb. per sq. in. as the steam is being used in an engine for a short time.

Determine what weight of water will be required in the storage tanks to supply sufficient steam to drive a turbine of 1000 H.P. for one hour with a consumption of 14 lb. of steam per H.P. hour under these circumstances.

The boilers work at a pressure of 150 lb. per sq. in. Find the state of the steam in the boilers necessary to restore the tanks to their initial condition after each period of delivery to the engine, and, assuming an exhaust pressure of 1 lb. per sq. in., determine what horse power would be developed by a Rankine engine using the same steam supplied direct to it by the boilers in the hour.

All radiation from the tanks may be neglected.

5. In a proposed cycle for a reciprocating steam engine, steam at a pressure of 150 lb. per sq. in. and superheated to 400° C. expands adiabatically in the cylinder to a pressure of 2 lb. per sq. in. at the end of the stroke: on the return stroke, condensation at constant pressure in the cylinder itself occurs for a period such that on adiabatic compression of the cylinder contents the steam will be just dry and at admission pressure, at the end of the stroke. This compressed steam is then, by means of suitable valves and an auxiliary pump, transferred to a superheater and replaced by superheated steam; expansion is then repeated. Thus the boiler is only required to superheat the steam.

Sketch the T-ϕ diagram for the cycle and, neglecting the work in auxiliaries, estimate its efficiency, comparing its value with that of the Rankine cycle between the same limits. Also find the relative volumes of the cylinders in the two cases to give the same power and the percentage clearance volume required in the proposed engine.

6. In a coal pulverising and drying plant of the ball mill type, 5 tons of coal are pulverised per hour with an expenditure of 20 H.P. for driving the mill. 6000 cu. ft. of dry air at 1200° C. enter the mill per hour and leave fully saturated with water vapour. The discharge temperature of the air, coal and water vapour is 50° C. and the pressure throughout is 14·7 lb. per sq. in., the barometric pressure being the same. The coal has a specific heat of 0·1 and enters the mill at 10° C.

Draw up a heat balance sheet on an hourly basis.

The mean specific heat of the air at constant pressure over the whole range of temperature may be taken as 0·245 TH.U. per lb. and R as 96 ft.-lb. per lb. The energy due to the change in pressure of the air may be neglected.

7. In the Atkinson cycle, for which $ABCD$ (fig. 1) is the indicator diagram, shew that the ideal efficiency of operation is

$$1-\frac{\gamma(\rho-r)}{\rho^{\gamma}-r^{\gamma}},$$

where ρ is the ratio of expansion and r the ratio of compression.

If the constant volume heating were extended to E, as shewn, so that the indicator diagram becomes $DEFBC$, the extra rise of pressure AE being xp, shew that the efficiency now becomes

$$1-\frac{\gamma\rho^{\gamma-1}(\rho-r)+x}{\rho^{\gamma-1}\{\rho^{\gamma}-r^{\gamma}+x\}},$$

where p is the induction pressure.

Expansion and compression are according to the law $pv^{\gamma}=$ const.

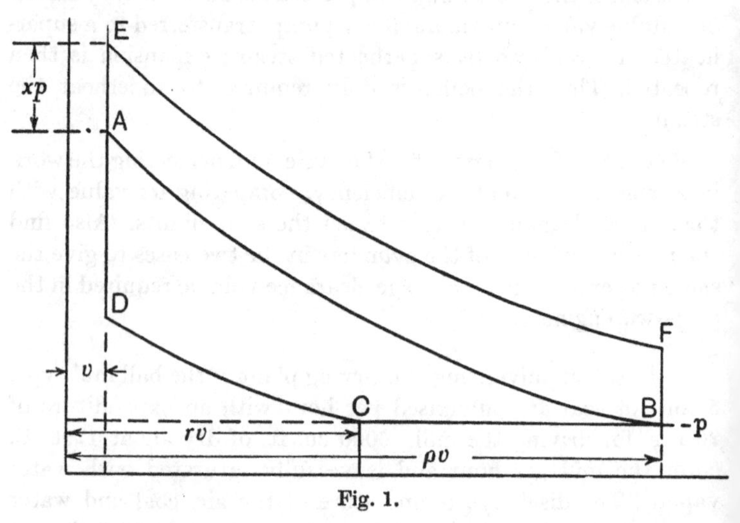

Fig. 1.

8. In a recent proposal for improving the efficiency of a triple expansion reciprocating steam engine the exhaust steam from the low pressure cylinder drives a turbo-electric generator, the electrical energy from which is used in part to drive auxiliary plant and in part to reheat the steam between the high and intermediate pressure cylinders.

In such a plant steam is supplied at 200 lb. per sq. in. and 250° C., it is expanded in the high pressure cylinder to 60 lb.

per sq. in., reheated at this pressure to 180° C. and expanded in the intermediate and low pressure cylinders to 4 lb. per sq. in. The condenser vacuum is 28 in. of mercury, and the atmospheric pressure is standard. Assuming ideal adiabatic expansion and neglecting all losses in the turbine and generator, find the power available for auxiliaries per lb. of feed water per minute.

Calculate the ideal overall efficiency of the plant and compare it with that of an engine working on the Rankine cycle between the same limits of temperature and pressure as the combined plant. Give a brief explanation of the difference between the two efficiencies.

9. Fig. 2 shews a developed section of a partial ring of nozzles for a pressure stage of an impulse turbine, the pressures on either side of the diaphragm being 120 and 40 lb. per sq. in.

Find the total throat area required to pass 3 lb. of steam per sec. The initial specific volume is 4 cu. ft. per lb., and owing to supersaturation the law of expansion may be taken as $pV^{1\cdot3} =$ const., and the critical pressure ratio as 0·545.

If the nozzles are of rectangular section throughout and of uniform radial height equal to the width of the throat, the number of nozzles 7 and the angle α of the jets 20°, find the length of the arc of admission. All dimensions to be taken on the centre line of nozzle.

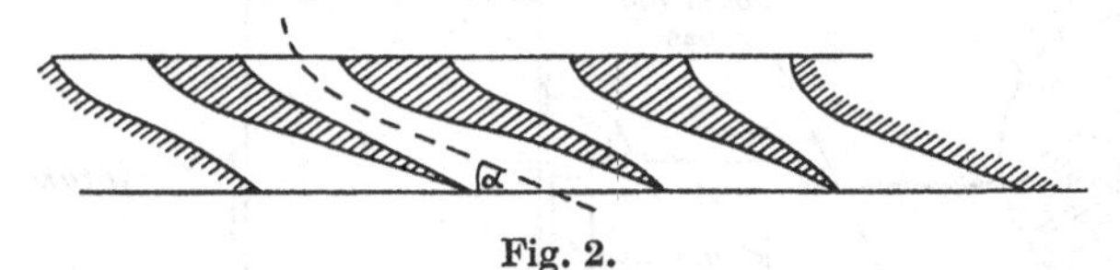

Fig. 2.

10. Test results are given below for a plant working on the Still system shewn diagrammatically in Fig. 3, in which the waste heat from an oil engine is applied to generate steam which is used on the lower side of the piston. The usual auxiliaries of a steam installation are included. No auxiliary oil was burnt under the boiler during the test.

Shew that the heat rejected from the internal combustion side is 57,750 TH.U. per min., and draw up a balance sheet indicating how this heat is disposed of (assume piston friction is one-half the total friction).

Data: I.H.P. oil: 1420.

I.H.P. steam: 240.

B.H.P.: 1520.

steam pressure: 120 lb. per sq. in. (assumed dry).

air pump discharge: 51 lb. per min. fed to boiler at 20° C.

condenser circulating water: 2400 lb. per min., inlet 8° C., outlet 18·5° C.

exhaust gases at chimney: 442 lb. per min. at 165° C. + 9·75 lb. steam formed during combustion.

engine room temperature: 14° C.

oil consumption: 8·3 lb. per min.

calorific value: 10,800 TH.U. per lb.

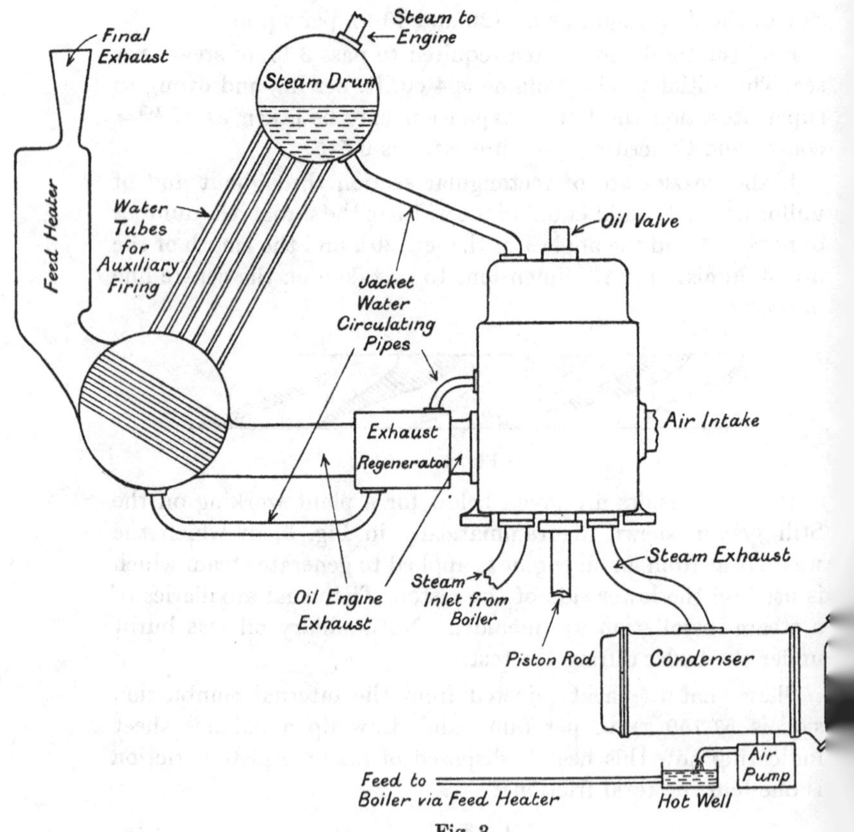

Fig. 3.

ANSWERS

PAPER I

1. 53·5 I.H.P.; 4·27 per cent.
2. 1400 TH.U.
6. 12·85 I.H.P.; 8·75 per cent.
7. 15 per cent.; 21·2 cu. ft.
8. 1409 ft. lb.

PAPER II

1. 62° C.
2. 18·7 lb.; 3520 cu. ft.
3. (i) 6·16 in.; (ii) 5·18 in.
4. 34·3 lb. per sq. in.; 62° C.
5.

Pressure, lb. per sq. in.	33·3	26·8	41·7
Temperature, ° C.	50	−13	131
Heat taken in, TH.U.	113·5	51·2	194·7

6. 0·242, 0·173.
7. 81 TH.U.

PAPER III

1. 153 cu. ft.; 7560 TH.U.
2. 2·49 cu. ft.; 17·3 TH.U.; −0·0625.

Pressure, lb. per sq. in.	100	25	14·37
Temperature, ° C.	1219	100	−59
Internal energy, TH.U.	207	17·3	−10·2
Entropy	0·274	0·0344	− 0·0625

4. 5·15 cu. ft.; 114·8° C.

PAPER IV

1. 23 TH.U.; 4·32 TH.U.; 18·7 per cent.
2. 4·42 per cent.
3. (1) 107·5 TH.U.; (2) 60 TH.U.
4. 14·5 per cent.; 12·73 TH.U.; 54 per cent.
5. (1) 17,500 ft. lb.; (2) 44·3 TH.U.; 31·8 TH.U.; (3) 28·2 per cent.
6. 16·5 TH.U.; 12·1 TH.U.
7. 0·46, −0·328, −0·132; 49·6 TH.U.; 15·3 per cent.
8. (1) 50 per cent.; (2) 4; 32·4 per cent.; 15·6 in.

PAPER V

1\. (1)

Pressure, lb. per sq. in.	100	50·2	10	19·9
Volume, cu. ft.	3·15	6·27	19·85	9·97
Q, TH.U.	22·4	0	−14·1	0
W, TH.U.	22·4	30	−14·1	−30
E, TH.U.	0	−30	0	30

(2)

Pressure, lb. per sq. in.	100	15·85	10	63
Volume, cu. ft.	3·15	19·85	19·85	3·15
Q, TH.U.	59·7	−29·8	−37·6	29·8
W, TH.U.	59·7	0	−37·6	0
E, TH.U.	0	−29·8	0	29·8

3\. 14,000 ft. lb.

4\. 11,200 ft. lb.; 97° C.

6\. 37·5 per cent.; 0·91.

8\. 1·416.

PAPER VI

1\. −0·079.

2\. 31·2 per cent.; 0·81.

3\. 20 H.P.; 45·5*d*, 1·8*d*.

4\. 36 per cent.; 0·0015.

5\. 42·3 per cent.

6\.

Source	Receiver	Working substance	Whole system
$-\frac{1}{473}$	0	$\frac{1}{443}$	$\frac{30}{443 \times 473}$
0	0	0	0
0	$\frac{1}{288}$	$-\frac{1}{318}$	$\frac{30}{288 \times 318}$
0	0	0	0

7\. 4·55 H.P.; 14·15 H.P.; 1·41 lb.; 2·69 lb.

8\. 2·33 H.P.; 6·39.

PAPER VII

49·5 per cent.

0·76; 223 TH.U.

Dryness: 1, 0·81, 0·62, 0·42.
Entropy: 1·609, 1·404, 1·183, 0·937.

84 per cent. dry; $7{\cdot}9 \times 10^6$ ft. lb. per min.

ΔE:	305,	-211,	-166,	72 TH.U.
ΔI:	336,	-232,	-182,	78 TH.U.
ΔQ:	336,	-211,	-182,	72 TH.U.

(1) 33,000 ft. lb.,	28 TH.U.,	78 per cent. dry.	
(2) 30,400 ,,	0 ,,	72·5 ,,	
(3) 30,800 ,,	10·7 ,,	75 ,,	

PAPER VIII

7·9 ft.

0·187; 416 TH.U.; 0·9 dryness.

73 lb.

5·36 TH.U.

(i) H.P. 100 to 23 lb. per sq. in.; L.P. 13·5 to 2 lb. per sq. in.; 9·5 lb. per sq. in.; 0·07.
(ii) 0·24 sq. in.

0·08 lb.; dry; $n=0{\cdot}71$; 11·45 TH.U.

56·4 and 15·65 per cent.; areas 1:1·05.

3 hr. 56 min.; 3 hr. 43 min.

PAPER IX

(i) 148 TH.U.; 70·5 per cent. dry.
(ii) 66 TH.U.; 0·255.
(iii) 0·5, 152° C.

(i) 14·51, (ii) 0·075.

(i) 158:147 TH.U.
(ii) H.P. 150 to 35 lb. per sq. in.; L.P. 15 to 2 lb. per sq. in.; 20 lb. per sq. in., 0·094.

(i) 0·0208; (ii) 0·25.

(i) 0·88, 182 TH.U.; (ii) 60 TH.U., 0·291; (iii) 170° C., 0·144.

65·76 TH.U.; 12·7 per cent.

375 lb.

(i) 65, 17·5, 4 lb. per sq. in., 0·926, 0·865, 0·81 dryness.
(ii) 56 TH.U., 0·275; (iii) 0·945 dryness, 0·28.

PAPER X

1. 47 per cent. dry; 2·1 cu. ft.
2. 110 lb. per sq. in.; 49 per cent. dry; 0·25.
3. 43·4 lb.
4. 52 TH.U., 13·3 TH.U., 65 TH.U., 67 TH.U.
5. (1) 60° C., 0·013; (2) 57·7° C., 0.
6. 62·8 per cent.
7. 15·75 lb.; 15·98 lb.
8. 5190 gallons.

PAPER XI

1. 77·5 lb.; 0·038 per cent.
2. 15 cu. ft.; 1·04 lb. per min.
3. 0·485, 0·383, 0·955, 0·242; 32 per cent.
4. 3·25 TH.U.
6. 109·6 lb.; 32·6 per cent.
7. (1) 84° C., (2) $2{\cdot}85 \times 10^5$ ft. lb., (3) 27·4 per cent.
8. $4{\cdot}03 \times 10^5$ TH.U., 31,250 lb.

PAPER XII

1. 34·7 per cent.
2. 57·85° C., 0·895, 560.
3. 141·3, 162 TH.U.; 34·9 per cent.; 147 TH.U.
4. 0·21 lb., 64·5, 76 per cent., 9·48 TH.U., 17·73 TH.U.
5. 77 TH.U.
6. 14·4 cu. ft.
7. $3{\cdot}35 \times 10^5$ TH.U.; 57 lb.
8. 82 TH.U. rejected.

PAPER XIII

1. 92 sq. ft.
3. 30·7 to 30·9 per cent.
4. 0·019 lb.; 0·328 lb.
5. 34·9 per cent.; 34·8 per cent.
6. 498 TH.U.
7. 1117 lb. per hour; 0·2 sq. in.
8. 30·8 to 32·2 per cent.

PAPER XIV

1. 16·7 per cent.; 43 lb.
2. 14·7 and 24·5 per cent.
3. 127 TH.U.; 21·35 per cent.
4. 52·8 I.H.P.; 12 per cent., 27·9 per cent.
5. 17·5 per cent.; 10·1 lb.; 23·3 per cent.
6. Heat: 0, −203·5, 0, 223 TH.U.
   Work: 25·1, 0, −27·6, 22·5 TH.U.
7. 0·158; 73·5 TH.U.; 5369 TH.U.; 13·6 per cent.
8. 83·5 per cent. dry; 2·8 TH.U.

PAPER XV

1. 132·6, 159 TH.U.
2. 134·2 TH.U. per 1 lb. of steam; 160 TH.U.
3. 160 I.H.P.; 17·15 per cent.
4. 144, 136, 88·2, 88·5 TH.U.; 8·61, 8·32 lb.
5. 2·23; 0·0575 lb.; 86·5 per cent. dry; 85, 78·4, 80·5 per cent.
6. 2 ft. diam., 4 ft. stroke; 62 R.P.M.
7. 7·1 per cent.
8. 81 per cent.

PAPER XVI

1. 0·0284 lb.
2. 21·4 in.
3. 65·2 per cent.; 76·8 per cent.
4. (1) 27·5 per cent.; 8·32 lb.; 30·3 per cent.
   (2) 28·2 per cent.; 7·45 lb.; 42·7 per cent.
5. 16·8 per cent.; 9·82 lb.; 24 per cent.; 140 TH.U.; 98·5 TH.U.
6. 2·58 cu. ft.; 15·3 cu. ft.; 37:5 cu. ft.
7. 0·1082, 25·2 per cent.; 0·0344, 33·3 per cent.; 5·8:1.
8. 38·6 TH.U.; 40 TH.U.

PAPER XVII

2. 82 per cent.; 18·8 per cent.
3. 129·6 TH.U. per 1 lb. of steam; 149 TH.U.
4. 161 TH.U., 26·1 per cent.; 135 TH.U., 183 TH.U., 24·3 per cent.
5. 102 TH.U.
6. 0·0587.
7. $k=\frac{1}{200}(T_1+T_2)-\frac{100}{T_1-T_2}(\phi_{s2}-\phi_{s1})$; 8·95 per cent.
8. (1) 29·15 per cent.; (2) 30·1 per cent.

PAPER XVIII

1. 173·2 TH.U.; 25·8 per cent.
2. 52·7 TH.U.; 80·3 TH.U., 150 TH.U.
3. (1) 128·1 TH.U.; (2) 127·8 TH.U.
4. 6·63 per cent.
5. 33·3 per cent.
7. 10·4 TH.U.; 26·5 to 27·5 per cent.
8. 1:2·4, 34·5 lb. per sq. in., 0·42 stroke, 76 per cent. dry, $n=1{\cdot}26$, 0·537 stroke, 0·8 per cent. dry.

PAPER XIX

1. (i) 18 cu. ft.; 0·145 lb. (ii) 43 cu. ft.; 0·222 lb.
3. 1520 cu. ft.
4. 28·34 in.
5. 5·766 TH.U.
6. 0·567 lb. per sq. in.
7. 1382 TH.U.; 91 lb. per sq. in.

PAPER XX

2. 154·8 lb. per sq. in.; 0·00515 lb., 175·1 TH.U.
3. 3·2 lb., 1932 TH.U., 1907·8 TH.U.
4. 65 cu. in.
6. 1825 cu. ft. per min., $6{\cdot}91\times10^4$ lb. per min.
7. 1385 cu. ft., $5{\cdot}54\times10^5$ lb. air, 4980 lb. of water make-up.

PAPER XXI

1. 37·5 I.H.P.; 69 per cent. dry.
2. 37·5 lb. per sq. in.; 421 ft. lb.; 6·07 I.H.P.
3. 73·5 I.H.P.; $pV^{1{\cdot}48}=540$, $pV^{1{\cdot}7}=220$.
4. 145 H.P.; 76 per cent.
5. 40·6 I.H.P.; 9·5 per cent.; 1371 gallons; 0·007 lb.
6. Stroke 30 in.; diam. 21·1 in.; 12 lb.; 210 per cent.
7. 5·1 cu. ft.
8. 37·5 in.; diam. 21·9 in.

PAPER XXII

1. 15·57 per cent.
2. 16 per cent.
3. 86·7 per cent.; 20·2 lb.; 98,000 ft. lb.; 148,500 ft. lb.
5. 406 TH.U.; 0·037 lb.
6. 346 I.H.P.; 85·2 per cent.; 17·2 per cent.; 4320 lb.
7. 77·4, 12·9, 86, 8·2, 2·86 per cent.
8. 80, 15·1, 86, 8·5 per cent.

PAPER XXIII

1. 0·413, 0·536, 600 TH.U.
2. 4 lb. per sq. in.
3. 206 lb. per sq. in.; 3775° C.
4. 8140 TH.U.
5. 2·416 lb., 10·4 lb.; 7653, 7420 TH.U.
6. 11·25 lb.; 15·42 lb.
7. 11 lb.; 8168 TH.U.; 15 lb.
8. (*a*) CO_2 13·06, O_2 2·04, N_2 84·9 per cent.;
   (*b*) CO_2 7·63, CO 5·08, O_2 4·53, N_2 82·76 per cent.

PAPER XXIV

1. 0·86 lb. per sq. in.
2. Equal proportions, 6 cu. ft.
3. 23·1 ft. lb.
4. 1·13 lb. per sq. in.
7. 0·875 cu. ft. per cu. ft.; CO_2 13·55, O_2 6·7, N_2 79·75 per cent.
8. 4·8 per cent.

PAPER XXV

1. 4·75:1; CO_2 5·5, O_2 10·5, N_2 84 per cent.; 0·97.
2. 8867 TH.U.; 26·8 lb.
3. (i) 10·62 lb., (ii) 22·6 lb., (iii) 0·455 lb.
4. Temperature: 109·4, 100·1, 26° C.
   Heat: 40, 43·6, 54 TH.U.
5. Gas to air by volume 1:10; 0·932.
6. 3360° C.
8. 21·3 lb.; 0·2392; 1610 TH.U.

PAPER XXVI

1. (1) 32·6 per cent., (2) 41·8 per cent.
2. 5·25; 48·5 per cent.; 15 per cent.; 1:3·2.
3. 32·8 per cent.; 26·24 per cent.
4. 56 per cent.; 67·7 per cent.; 582 TH.U.
6. 0·08 lb.; 724 lb. per sq. in. and 2463° C.; 61 lb. per sq. in. and 1107° C.
7. 41·5 H.P.
8. 36 per cent.; 27·6 per cent.; 49·9 per cent.

PAPER XXVII

1. 7·15; 54·5 per cent.
2. 0·425 lb. per B.H.P. hour.
4. 24 I.H.P., 35·5 per cent., 47·1 per cent.
5. 168° C.
6. 23·2 H.P.
7. 0·123 lb.; 173 lb. per sq. in.; 360° C.; 526 lb. per sq. in.; 1660° C.; 51·1 per cent.
8. 0·00306 lb., 60° C., −56·5° C., 2873 ft. lb.

PAPER XXVIII

1. (1) 5·25, (2) 78·75 lb., (3) 83·3 per cent., (4) 25·2 per cent., (5) 1·05 H.P.; 5·4 per cent.
2. 1·1 cu. ft.
3. 44° C., 241° C.; 97·4 lb. per sq. in.
4. 0·842; 30 I.H.P.
5. 19·6 I.H.P.
7. 102° C.
8. 2·6 TH.U.

PAPER XXIX

1. 285 ft. lb.; 14·9° C.
2. 85·5 per cent.; 24·6 per cent.
4. 96·8 TH.U.
5. 165° C.
7. 1·33.

PAPER XXX

1. 1·94 TH.U.; 141 lb. per sq. in.
2. 78° C.; 0·00651 lb.
3. 81° C.; 2° C.
8. 102° C.; 5·4 TH.U.

PAPER XXXI

1. 55·5° C.; 8082 TH.U.
2. 204 TH.U.; 7·055 lb.; 4164 TH.U.
3. 8280 TH.U.; 68·5 per cent.; 18·5 per cent
4. 74·8 per cent.; 212,000 TH.U.
6. (*a*) 30·2 per cent.; 15·82 per cent.
7. 77·1 per cent., 15·95 per cent.
8. Inconsistent.

PAPER XXXII

1. (1) 147, (2) 143, (3) 15·1 lb. per sq. in.
2. 340° C.; 2·89 st. cu. ft.; 67·4 H.P.
3. $7{\cdot}97 \times 10^4$ ft. lb., $4{\cdot}39 \times 10^4$ ft. lb.; 45 per cent.
4. $\left\{\dfrac{(1+k)\,\alpha^{\frac{1}{n}} + m x^{\frac{1}{n}}}{k+m}\right\}^n p_0.$
5. $\sqrt{p_1 p_2}$, $\sqrt[3]{p_1^2 p_2}$, $\sqrt[3]{p_1 p_2^2}$. 9×10^4 ft. lb., $7{\cdot}6 \times 10^4$ ft. lb., $7{\cdot}2 \times 10^4$ ft. lb.; 27 TH.U., 55 TH.U.
7. 1·15.
8. 394 H.P.; 35 per cent.

PAPER XXXIII

1. 35·7 lb. per sq. in.; 312 ft. lb.; 0·0112 lb.
2. (i) 527·5 TH.U., (ii) 51·9 per cent.
3. 186·6 lb. per hr.; 0·202 cu. ft.; 77·9 lb. per hr.
4. 0·863.
5. 880 H.P., 585 H.P.; 1·5.
6. 0·8 ft.; 3·92, 1·93, 0·96 in. diam.
7. $5{\cdot}67 \times 10^4$ ft. lb.
8. 4·2 per cent. of stroke.

PAPER XXXIV

1. 25·5 TH.U.
2. (1) 5·4, (2) 3·59.
3. 3·13; 82° C.; 2·38.
4. (1) 3·89, (2) 2·66.
   (1) 8·6, (2) 7·37.
5. 2·04.
6. 22·3 TH.U.
7. 9·05 H.P.; 3 cu. ft.
8. 1·24 H.P.

PAPER XXXV

1. 66·5 H.P.; 50° C.; 63·5 H.P.
2. 2250 lb.; 18·3 H.P.
3. 244 TH.U.; 57 cu. ft.; 4 per cent.
4. NH_3 (1) (*a*) 10·8, (*b*) 90 per cent. dry; (2) 260 TH.U., (3) 3·85 lb., (4) 44·4 cu. ft., (5) (*a*) $6{\cdot}28 \times 10^4$ ft. lb., (*b*) $2{\cdot}42 \times 10^5$ ft. lb., (6) 5·7.
   SO_2 (1) (*a*) 13·25, (*b*) 90·5 per cent. dry, (2) 74 TH.U., (3) 13·5 lb., (4) 123 cu. ft., (5) (*a*) $1{\cdot}8 \times 10^4$ ft. lb., (*b*) $2{\cdot}43 \times 10^5$ ft. lb., (6) 5·6.
5. 85 TH.U.; 4·78.
6. (1) 3·8, (2) 1·47 H.P.; 3·1; 1·8 H.P.
7. 5·83; 19·55 H.P.; at 92° C.
8. (*a*) 5·11, (*b*) 78·1 per cent. dry, (*c*) 12·9.

PAPER XXXVI

1. 0·029 sq. in.
2. 100·5° C.
3. 14·8 per cent.; $n = 1{\cdot}113$; 15·1 TH.U.
4. 3210 ft. per sec.; 86·5 per cent. dry; 0·056 lb.
5. 981 lb. per hour.
6. 1315 ft. per sec.; 0·949.
7. 0·0472 sq. in. throat; 0·514 sq. in. opening.
8. 41 sq. in.

PAPER XXXVII

1. $24^\circ \cdot 52'$; $7 \cdot 7 \times 10^4$ ft. lb.
2. $29^\circ \cdot 32'$; 80·5 per cent.
3. 3780 ft. per sec.; $29^\circ \cdot 15'$; $1 \cdot 78 \times 10^4$ ft. lb.
4. 18·5 per cent.
5. $2 \cdot 94 \times 10^4$ ft. lb.; 7·7 lb.; 3800 ft. lb.
6. 0·426 in.
7. (i) 26·8 ft. per sec., (ii) 2930 ft. lb., (iii) 69 per cent.
8. 38·4 TH.U.; 14·6 TH.U.; 0·83 TH.U.

PAPER XXXVIII

1. 17·28 TH.U.; 17·28 and 21·6 TH.U.; 35 lb. per sq. in., 92·7 per cent. dry.
2. 13·72 per cent.
3. 35·6 TH.U.; 4·38 lb.; 11·82 TH.U.
4. 36·4 per cent., 5·3 lb.; 40·2 per cent., 5·72 lb., 21·2 per cent.
7. (*a*) 12·74, (*b*) 12·1 per cent., 35·5, 27·1 per cent.
8. 170, 120 lb. per sq. in.; 205° C.

PAPER XXXIX

1. 42·9 TH.U.
2. 111 TH.U.; 4·65° C.
3. 0·0118 B.TH.U.; 0·1835 H.P.
4. 27·5 TH.U.; 16·8 TH.U.
6. 0·00363 H.P.

PAPER XL

1. $10 + 15 \cdot 12x - 1 \cdot 412x^2$.
2. 122° C.; 7·95 TH.U.
3. (1) 57 per cent., (2) 91·5 per cent., (3) 93·5 per cent.
7. 0·12 in.; 0·038° C.

PAPER XLI

2. 295 TH.U.
3. 15·15 per cent.; 21·2 per cent.
5. (1) 92·1 TH.U., (2) 84·1 TH.U., (3) 198·7 TH.U.; 31·9 per cent.; 5·42 lb.
6. 23 per cent., 20·2 per cent.
7. 6·87 lb.
8. 11:12.
9. 6·62 TH.U.; 154·5° C.
10. 2·14, 2·9.

PAPER XLII

3. $2{\cdot}68 \times 10^4$ ft. lb.; 33 TH.U.
4. 148·5 cu. ft.
5. CO 39·8, N 43·4, H 16·8 per cent.
6. 86° C.
7. 0·0094, 0·0054; 15 TH.U.
8. $1{\cdot}36 \times 10^5$, $1{\cdot}89 \times 10^5$, $2{\cdot}55 \times 10^5$ ft. lb.; (*a*) 4026, (*b*) 3720 ft. per sec.
9. 1062 ft. per sec.
10. 470 lb. per sq. in.; 39·5 per cent.

PAPER XLIII

4. 485 lb.
5. 95·5 per cent. dry; 4·8 per cent.
6. 133·2 lb. per sq. in.
7. 1·55 lb.; 0·48.
8. 156 lb. per sq. in.
9. NH_3 8·4; CO_2 5·85.
10. 1·775 in.

PAPER XLIV

1. 17·4° C.
2. 1·404; 96·6 per cent.; 0·046.
5. 96° C., 99° C.
7. 1·935.
8. 208° C.; 0·005.
9. 24°·49′, 31°·7′, 39°·40′; 307 ft. per sec. at 90°; $5{\cdot}63 \times 10^4$ ft. lb.; 1:1·39:1·77:2·45.

PAPER XLV

2. 82·1 TH.U.
3. 66·5° C.
4. 3750 lb.; $3{\cdot}64 \times 10^5$ ft. lb., $5{\cdot}74 \times 10^5$ ft. lb.
5. (1) 0·38 lb.; 97·5 per cent. dry, (2) 92·9 per cent. dry.
6. 2042° C.; work done 8900, energy loss 9000 ft. lb.
8. 2·27:1.
9. 87 per cent. dry; 1500 ft. per sec.; 7 per cent.
10. 23°·15′, 29°·8′, 41°·40′, 498 ft. per sec.

PAPER XLVI

1. −66·5° C.; 37·43 cu. ft.
2. 29·2 per cent.
3. $3\cdot7 \times 10^4$ cu. ft.
4. 68° C.
6. 0·4, 2·16, 286 TH.U. per min.
7. NH_3: 255 TH.U.; 4·8;
   SO_2: 73·2 TH.U.; 4·85.
8. 2·8 TH.U.; 2270 ft. per sec.
9. 173 TH.U.; 390 cu. ft. per lb.; 734 ft. per sec.; 3·47 per cent.
10. 898 TH.U.

PAPER XLVII

1. (*a*) 1500° C. abs.; 1500 lb. per sq. in.; (*b*) 1525° C.; 15·25 lb. per sq. in.
3. (i) 15 to 2 lb. per sq. in.; 22 per cent.; 0·07.
   (ii) 0·116; 0·942 at 35 lb. per sq. in.; 0·895 at 6·5 lb. per sq. in.
4. 9·14:10.
5. 5:4.
6. 11·3 per cent.
7. 66·4 per cent.; 12 lb.; 22·8 per cent.; 8280 TH.U.
8. 5·95; 7·3.
10. 150 lb. per sq. in., 300° C.; 90 lb. per sq. in., 257° C.; 55 lb. per sq. in., 215° C.; 30 lb. per sq. in., 170° C.; 1·759.

PAPER XLVIII

1. 115° C.
2. 0·813; 8 TH.U.; 32 TH.U.
3. 24·7 per cent.; 25·7 per cent.; 0·705.
6. (*a*) 0·0291 lb.; 0·01245 lb.; (*b*) 294° C.; 0·00115 lb.
7.

Pressure, lb. per sq. in.	25	50	75	100
True eff.	0·821	0·809	0·79	0·765
Apparent eff.	0·967	0·938	0·91	0·883

8. 162° C. super heat; 1462 ft. per sec.; 1·815 sq. in.
9. 55·1 TH.U.; 71·3 per cent.
10. (i) 32·6 per cent.; (ii) 34·94 per cent.

PAPER XLIX

1. 8·2 TH.U. rejected.
2. (1) 27·1 per cent., (2) 5·31 lb., (3) 79·3, 94·1 per cent., (4) 30·6 per cent.
8. 51·8 cu. ft.
9. (1) 6·32, (2) 3·66.
10. (1) (*a*) 1122, (*b*) 1053, (2) 0·445, (3) 11 TH.U., (4) 0·785, (5) 1·48.

PAPER L

1. 0·0027.
2. 91·3 TH.U.
4. $2{\cdot}25 \times 10^5$ lb.; 99·3 per cent. dry; 1750 H.P.
5. 41 per cent., 28·1 per cent.; 4·35:1; 1·89 per cent.
6. TH.U. 26,400 air — 1,100 air
   11,200 coal — 56,000 coal
   28,280 work — 4,100 vapour.
8. 15 TH.U.; 30·1 per cent., 30·3 per cent.
9. 1·75 sq. in.; 11·4 in.
10.

Oil	89,600	Steam	32,800	I.H.P.	5,660
Friction	1,650	Exhaust	22,410	Condenser	25,200
	91,250	Balance	2,540	Balance	1,940
I.H.P.	33,500				
	57,750 TH.U.				

Printed in the United States
By Bookmasters